Virtual Reality Headsets – A Theoretical and Pragmatic Approach

T0179408

Virtual Reality Headsets –
A Theoretical and
Pragmatic Approach

Philippe Fuchs
Ecole des Mines, ParisTech, Paris, France

Co-authors:
Judith Guez, Olivier Hugues, Jean-François Jégo, Andras Kemeny & Daniel Mestre

CRC Press
Taylor & Francis Group
Boca Raton London New York

CRC Press is an imprint of the
Taylor & Francis Group, an **informa** business

A BALKEMA BOOK

CRC Press
Taylor & Francis Group
6000 Broken Sound Parkway NW, Suite 300
Boca Raton, FL 33487-2742

First issued in paperback 2019

© 2017 by Taylor & Francis Group, LLC
CRC Press is an imprint of Taylor & Francis Group, an Informa business

Originally published in French as: "Les casques de réalité virtuelle et de jeux vidéo",
© 2016 Presses des Mines, Paris, France

Typeset by MPS Limited, Chennai, India

No claim to original U.S. Government works

ISBN-13: 978-1-138-63235-6 (hbk)
ISBN-13: 978-0-367-88835-0 (pbk)

This book contains information obtained from authentic and highly regarded sources. Reasonable efforts have been made to publish reliable data and information, but the author and publisher cannot assume responsibility for the validity of all materials or the consequences of their use. The authors and publishers have attempted to trace the copyright holders of all material reproduced in this publication and apologize to copyright holders if permission to publish in this form has not been obtained. If any copyright material has not been acknowledged please write and let us know so we may rectify in any future reprint.

Except as permitted under U.S. Copyright Law, no part of this book may be reprinted, reproduced, transmitted, or utilized in any form by any electronic, mechanical, or other means, now known or hereafter invented, including photocopying, microfilming, and recording, or in any information storage or retrieval system, without written permission from the publishers.

For permission to photocopy or use material electronically from this work, please access www.copyright.com (http://www.copyright.com/) or contact the Copyright Clearance Center, Inc. (CCC), 222 Rosewood Drive, Danvers, MA 01923, 978-750-8400. CCC is a not-for-profit organization that provides licenses and registration for a variety of users. For organizations that have been granted a photocopy license by the CCC, a separate system of payment has been arranged.

Trademark Notice: Product or corporate names may be trademarks or registered trademarks, and are used only for identification and explanation without intent to infringe.

Library of Congress Cataloging-in-Publication Data

Visit the Taylor & Francis Web site at
http://www.taylorandfrancis.com

and the CRC Press Web site at
http://www.crcpress.com

Table of contents

Preface

Since the beginning of 2014, a lot of companies have been announcing new products in the field of Virtual Reality (VR). Oculus Rift, Hololens, Kinect, Leap Motion, Magic Leap ... so many new names that offer new opportunities. Whatever their nature (VR headsets, sensors, glasses ...), these new materials have a common denominator: they cost much less than the previous equipment and therefore enable consideration of applications for the general public.

These announcements have been relayed by a large quantity of news items in the media but, unfortunately, most of them offer an incomplete or false analysis, because of a misunderstanding of VR. Especially because many of these "journalists" are discovering an area that is far from emerging. VR has existed for over a quarter of a century with its uses, its companies and research laboratories. From all applications used for a long time: automotive construction, conception of buildings, design and learning of surgical procedures, archaeological reconstructions, industrial and sport training, etc.

The most common error is the confusion between VR and Immersion, allowing the user to "plunge" into a virtual environment by providing his brain artificial sensory stimuli. Immersion is necessary for VR, but it is not sufficient. The second essential pillar of VR is Interaction between the system and the user so that he can perform a task in a virtual environment. Thus, one cannot talk about VR for an exploration of a site via a 360° picture using a VR headset (Head Mounted Display) or for a movie, even if the image is projected in relief on a large screen, accompanied by a spatial sound, because the viewer is fully passive.

Much more serious than this terminological confusion is that professionals, in both the corporate and academic world, have known for many years that a VR immersion causes problems that must not be overlooked. From a feeling of discomfort to a faintness, VR sickness (cybersickness) has been studied for a long time to try to understand its causes and to mitigate its effects. Curiously, no trace of these problems can be found in these recent articles; no study is carried out by the VR headsets manufacturers who simply publish a license dismissing any responsibility for any problem. Illustration: in the instructions supplied with a famous recent VR headset, the manufacturer advises to not drive a vehicle just after using the product! The desire to protect oneself against any legal risk is not sufficient to justify such a recommendation. In practice, professionals limit the duration of use of VR headsets for the reasons mentioned above. What will happen if our children play an immersive game several hours daily? At an age when their visual system is still developing, what consequences are to be feared? It is really surprising that at a time when the precautionary principle is needed in many areas, this question does not raise more interest, nor, to our knowledge, in long-term studies involving experts from various domains.

It is therefore essential to present, as clearly as possible, VR concepts, to explain the operating principles of these new VR headsets and to study their use for understanding both the opportunities and the risks. Unfortunately, there was a missing link between these articles for the general public and communications in journals or at scientific conferences for experts, whether industrial or academic.

It is to fill this gap that Philippe Fuchs has decided to write this book. It fully describes the concepts, modes, uses and ways to avoid discomfort and possible faintness. I am betting that this book will meet with success for several reasons: first, Philippe Fuchs is a pioneer of VR in France and he has been helped by experts to write this book. Second, because he has already demonstrated the qualities necessary for such success with the "Traité de la réalité virtuelle" ("Treatise on Virtual Reality" in French with an abbreviated English edition: "Virtual Reality: Concepts and Technologies", published by CRC Press). This collective work was supervised by Philippe Fuchs. Recall that there is no equivalent of this encyclopedia whose latest edition, over 2000 pages in length, brought together 101 authors!

This is no surprise at all, coming from a man who likes to take on the wildest challenges, not just academic, like running seventy kilometers a day, six days a week, for five months, linking Paris to Beijing for the Olympics opening in 2008! The question I ask myself now is: what will be his next challenge?

Pascal Guitton
Professor of Computer Science at the University of Bordeaux
Founding Member and Président of the French National Association
of Virtual Reality (2009–2011)
Scientific Director of Inria (2010–2014)

About the author

Philippe Fuchs, PhD, Professor in Mines ParisTech engineering school (Paris), PSL Research University, is the leader of the "Virtual Reality & Augmented Reality" team. The field of his research is the theoretical approach of VR and industrial applications. The team's lines of research focus mainly on human "behavioural interfacing" in a virtual (or mixed real/virtual) world, by making judicious use of a person's natural behaviour on sensory motor and mental levels. Our methodology for designing a VR system has been extended on the technical and psychological levels thanks to collaboration undertaken this year with ergonomists and psychologists. The team's lines of research focus on visual interfaces, especially VR headsets and stereoscopic vision.

- Email address: philippe.fuchs@mines-paristech.fr
- www.philippe-fuchs.fr

Co-authors

Judith Guez is a digital artist and researcher. She creates experiences, art installations and performances using Virtual Reality. After graduating with a Master of contemporary arts of the University of Paris 8, Judith obtained a PhD in digital arts at the University Paris 8 in France. She teaches Virtual Reality at the Art department of the Paris 8 University and is also a member of the INREV virtual reality laboratory at the Paris 8 University

Olivier Hugues received his PhD in computer science from the University of Bordeaux I (Aquitaine, France). He then worked at the Université du Québec in Abitibi-Témiscamingue (QC, Canada) in the mining sector and, subsequently, served as an Assistant Professor in MINES ParisTech's Center for Robotics in the "Virtual Reality & Augmented Reality" team. His interest fields include the understanding of augmented reality from the user point of view and human-robot collaboration.

Jean-François Jégo is a digital artist and researcher. He creates immersive and interactive experiences, art installations and performances using Virtual Reality. After graduating with a Master of contemporary arts of the University of Paris 8, Jean-François obtained a PhD in computer science and virtual reality at the CAOR robotics center of Mines ParisTech in France. He teaches Virtual Reality at the Art department of the Paris 8 University and he is also a member of the INREV virtual reality laboratory at the Paris 8 University.

Andras Kemeny, Doctor ès Sciences (1983, Université Paris Est), is the Head of the Virtual Reality and Immersive Simulation Center of Renault-Nissan Alliance Engineering, Digital Vehicle Testing and Customer Performance Division and Director of the Laboratory of Immersive Visualization (LIV), a joint Renault – Arts et Métiers ParisTech research laboratory.

Daniel Mestre, Ph.D, is a research director at Centre National de la Recherche Scientifique (CNRS) and affiliated to the Institute of Movement Sciences (CNRS and University Aix-Marseille II). He is also the head of the Mediterranean Virtual Reality Center (CRVM) and founding member of the French Association for Virtual Reality (AFRV).

Part I

A theoretical and pragmatic approach for VR headsets

Chapter 1

Introduction and challenges

Virtual reality has developed in the world over the last twenty years. It potentially opens up new perspectives for our society. But let's be realistic – first of all, virtual reality creates many scientific challenges for researchers and professionals. Being aware of the immensity of the task at hand, we have participated enthusiastically in helping virtual reality in France blossom. For our part, we conducted theoretical and applied research on the interface of the subject (user interface) in a virtual environment. However, no researcher can ever have a precise, essentially interdisciplinary knowledge of all the sectors of virtual reality. We wanted other French researchers, Pascal Guitton and Guillaume Moreau, to participate in writing the book "Virtual Reality: Concepts and Technologies", publisher CRC Press, originally published in French as: "Le Traité de la Réalité Virtuelle", with 101 authors, five volumes, publisher "Les Presses des Mines".

Thanks to technological developments, man has been able to satisfy this need through various but set representations of the world that are mainly audio or visual. Set in the sense that the user can observe the representation only as a *spectator*, be it a painting, a photograph, or a film of real or computer-generated images. Virtual reality offers him an additional dimension by providing him a virtual environment in which he becomes the *actor*. The readers must not be mistaken; the novelty is not in the creation of virtual environments that are increasingly efficient in terms of their representation, but rather in the possibility of being able to "act virtually" in an artificial world (or "interact" in a more technical sense). Imagining virtual reality has been possible only recently, thanks to a significant increase in the intrinsic power of computers, especially the possibility of creating computer-generated images in real time and enabling a real-time interaction between the user and the virtual world.

And 2016 is the year of virtual reality in the field of general public applications: the new VR headsets (or Head Mounted Display – HMD) are sold for the general public with a low cost price. The reader must note that it is these technical developments that have permitted the new boom of virtual reality. The technology is more affordable and more accessible now. In November 2015, I decided to write a new French book about the use of this new visual interface: VR headset, originally published in French as: "Les casques de réalité virtuelle et de jeux vidéo", publisher "Les Presses des Mines", May 2016.

The user will no longer experience virtual worlds on simple screens, but through VR headsets offering new possibilities of visual immersion, the advantages and limitations of which are the primary concern of the present work. This does not mean that industries interested in using VR within their sectors should automatically choose VR headsets when other more suitable visual interfaces exist. This question too will be addressed here.

Though the term "virtual reality" generally refers to professional applications built around interactive environments, VR researchers and manufacturers are well aware that the same concepts and core technologies, whether hardware or software, are used in the video game industry. Examples of this abound, with such VR software as Virtools and Unity, developed primarily for video games, being used for professional VR applications, both in the past, in the case of Virtools, and the present, in that of Unity. As for hardware, Microsoft's input motion control device for Xbox 360 consoles, Kinect, was rapidly diverted from its original use to be utilised by researchers today for a whole host of non-gaming VR applications. Needless to say, such technological strides are owed solely to the enormous wealth of the video game industry, which can afford to pour huge sums of money into the development of pioneering technology. There is no need to go on about the enormous financial interests at stake—just look at the stir caused by Facebook with its purchase of Oculus!

We have been using the term "virtual reality" for more than twenty years. This term is debatable and has been questioned by some. The oxymoronic expression *virtual reality* was introduced in the United States by Jaron Lanier in the 1980s.

After an initial reassessment of what virtual reality actually is, we will need to address a number of issues regarding human sensorimotor response, in both the real world and in virtual environments. A particular focus will be placed on the sense of vision, given the *visually invasive* nature of VR applications, which directly impact both the sensory perception and the motor response of users. Furthermore, a refresher course on the five senses—which in actual fact are more than five!—will be provided for those readers who may not be overly familiar with the topic, myself included at the start of my career. This review of sensory fundamentals will be enormously helpful in understanding the problems and the solutions particular to VR headset usability. In short, to fully exploit the capabilities of VR headsets, a good knowledge of human vision is necessary.

Virtual reality holds a special position in the usual scientific scheme by coupling human sciences with engineering. This position is an advantage of the intrinsic interdisciplinary nature of this domain. However, this position is also a difficulty to overcome, on the one hand in terms of training the actors of the domain, and on the other hand in terms of recognition for this multidisciplinary foundation on the part of the various disciplines that enrich it. For example, it would be too simplistic to consider virtual reality merely as a branch of computer science. Though computers make it possible to effectively program and simulate the virtual worlds, interaction of man with these worlds is possible only through software programs and technical devices compatible with cognitive, perceptive and social processes. Conversely, better understanding and formalising of the difficulties and characteristics of cognition and interaction in the virtual worlds offers an empirical foundation to stimulate research and innovation.

In any virtual reality application, the person is immersed in and interacting with a virtual environment. He perceives, decides and acts in this environment, a process schematised in a standard "perception, decision, action" loop, which must be achieved within the technical, physiological and cognitive constraints (Figure 1.1).

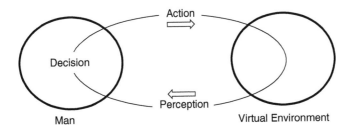

Figure 1.1 The standard "perception, decision, action" loop.

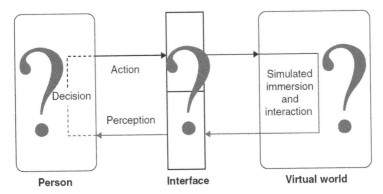

Figure 1.2 Diagram showing the issues of virtual reality, based on the "perception, decision, action" loop.

Three fundamental issues of virtual reality can be deduced from this diagram (Figure 1.2):

• The issue of the analysing and modelling the human activity in real environment and in virtual environment;
• The issue of analysing, modelling and creating an interface for the subject for his immersion and interaction in a virtual environment;
• The issue of modelling and creating the virtual environment.

VR headset utilisation raises the three following questions:

– How are VR headset users affected by display latency and sensorimotor discrepancy in virtual worlds?
– Which types of user interfaces and interactions generate the above latencies and discrepancies?
– How can various systems and computer algorithms help reduce the effects of the aforesaid artefacts?

The book is divided into two sections: "Theoretical and pragmatic approach for VR headsets" and "VR headset applications". The chapter 2 introduction to virtual reality clarifies the book's scope and presents the theoretical approach, also known as "$3I^2$ model". Then, chapter 3 about human senses is necessary to understand the sensorimotor immersion, especially vision in real and virtual environments. These chapters are followed by chapter 4 which presents the different visual interfaces available, chapter 5 is about the VR headsets, chapter 6 about the user interfaces exploited with VR headsets and chapter 7 about the commercial devices available on the market (author: Olivier Hugues). These visual interfaces can imply comfort and health problems with the sensorimotor discrepancies. Chapter 8 is devoted to these problems, followed by chapter 9 that gives a detailed discussion of methods and 32 solutions to remove, or to decrease the VR sickness. Then, the "VR headset applications" second section presents different VR applications that use headsets (Behavioural Lab Experiments – author: Daniel Mestre; Industrial uses of VR headsets – author: Andras Kemeny; Creating Digital Art Installations With VR Headsets – authors: Judith Guez and Jean-François Jégo) and the last chapter gives conclusions and future VR challenges.

We can only wonder whether developers are fully aware of the enormity of the task at hand, of the obstacles that need to be overcome and the laws that need to be complied with for designing effective VR applications. Moreover, if these issues are indeed resolved, can we be certain that users will be able to successfully adapt to visual immersion? The main purpose of this work is to present a general review of such questions, keeping in mind that VR technologies "perturb" our physiological and sensorimotor response. This can be likened to the sensorimotor discrepancy created by the accommodation/vergence conflict of stereoscopic vision. However, professionals in the field, or stereographers, have long been aware of the rules governing stereoscopic image making (3D). The first of these, dating back more than 150 years, was determined by David Brewster in 1856, and many experiments have since been conducted to establish the types of imagery that viewers are comfortable with. As for VR headsets, the latest of these visually intrusive interfaces, user feedback is still very limited. We can only hope that the experts will address this question to which very little attention has been paid to date. In part to protect themselves from this lack of scientific data, some VR headset designers warn of the potential risks linked to their products and have set a minimum user age. At first glance, these issues seem far more complex than those of stereoscopic vision, from a purely intuitive perspective at any rate. Furthermore, since VR headsets are not yet used by the general public, researchers lack data from large target group studies. The fact is, it remains to be seen how well we will adapt to visual immersion in virtual worlds; some people may be more sensitive than others yet without really knowing why. Another issue raised by VR headset use is whether the risk of video game addiction will be greater for gamers wearing VR headsets than for those interacting via a simple screen. This question, which is a matter for psychologists and psychiatrists, will not be discussed here, though, to my knowledge, no serious research has been conducted on the subject to date.

Though the effects of VR headset use raise special problems of comfort and health, they also afford exciting possibilities, as virtuality allows us to go beyond the limits of reality. Questions of physical perception, of experiencing "presence" in a space in which we are either unable, or only partially able, to see our own body (our avatar's hands) must also be answered. How do we respond to seeing ourselves visually but in another body, to what could be qualified as an out-of-body experience? All of these behavioural questions, as well as our capacity to respond to potential virtual environments that have little in common with real-world environments, need to be addressed by researchers in the cognitive sciences.

So the question is – how can we take advantage of this ground-breaking visual interface while minimising its drawbacks? This work proposes to provide insights on the development of effective well-thought out VR headsets, as it is essential, in my view, to inform both professionals and the general public about this technical innovation that is anything but trifling.

Chapter 2

Concepts of virtual reality

2.1 DEFINITIONS OF VIRTUAL REALITY

Defining virtual reality is an indispensable task. In literature, we still find definitions that inappropriately mix the purpose of virtual reality, its functions, applications and the techniques on which it is based. We must reject these approaches, firstly because they are centred on only one particular technology, and secondly because they are extremely restrictive in terms of scientific issues related to the complexity of the dimensions involved in the interaction between the human user and the virtual environments. We have given definitions with various levels to give a clear picture of the domain of virtual reality.

2.1.1 Purpose of virtual reality

Before elaborating its functions or techniques, it would be wise to first determine the purpose of virtual reality shared by all French practitioners. After having studied the objective that is common to all its applications, we can claim that (Fuchs, 1996; Arnaldi & Fuchs, 2003):

> The purpose of virtual reality is to make possible a **sensorimotor** and cognitive activity for a person (or persons) in a digitally created artificial world, which can be imaginary, symbolic or a simulation of certain aspects of the real world.

A simulation of certain aspects of the real world: These aspects are to be determined at the time of designing the application. You will realise that this initial phase of designing is fundamental and thus must be analysed clearly. Errors, which are found often, are of the designer who tries to reach the highest "degree of realism". This incorrect approach is taken without trying to understand precisely which aspects of reality are necessary to be covered in the application. It is completely absurd to naïvely expect, if possible, that the behaviour of the virtual world would be exactly identical to that of the real world. If we want to create a "virtual" reality, modifying the aspects of the "real" reality is well within its purpose. For instance, it can be used for training in a virtual environment, without real danger for the trainee.

A symbolic world: We can also use symbolic representations to improve the understanding of the simulated world. Virtual reality is then used either to represent a phenomenon (structure of molecules, flow of fluids, etc.) or to add symbolic concepts to the simulated real world. These concepts help the user to have a better mental representation of his environment. For example: information can be displayed in diagrams to help the user understand the structure of a mechanism or the plan of a task to be completed.

An imaginary world: Virtuality is used to create an unreal world, a figment of imagination of an artist or a science-fiction writer. In this case, the created world does not have to be a simulation of the real world.

There are certainly different ways of using the potentials of virtual reality; the three cases can obviously be associated to a single application. A sensorimotor activity for a gamer is low when he plays video games with a computer and a single screen. But the sensorimotor activity will be greater with video games which use VR headsets and hand tracking.

2.1.2 Functional definition

In 1996, Fuchs proposed a taxonomy based on "theoretical" functions: Vis-à-vis his own perception of the reality, man has conceptualised the notions of time and space on which he can interact only as per the immutable physical laws.

> *Virtual reality will help him to come out of the physical reality to virtually change* **time, place** *and (or) the type of* **interaction:** *interaction with an environment simulating the reality or interaction with an imaginary or symbolic world.*

This definition refers to the opposite demand of the authors of tragedies of the 17th century, who advocated the rule of three units – time, place and action. See this approach in the article (Nannipieri & Fuchs, 2009).

2.1.3 Technical definition

A more technical and literal definition of virtual reality attempts to characterise the domain. Immersion and interaction are the two key words of virtual reality. The technical definition of virtual reality is (Arnaldi & Fuchs, 2003):

> *Virtual reality is a scientific and technical domain that uses* **computer science (1)** *and* **behavioural interfaces (2)** *to simulate* **in a virtual world (3)** *the behaviour of 3D entities, which* **interact in real time (4)** *with each other and with one or more users in* **pseudo-natural immersion (5)** *via sensorimotor channels.*

This definition introduces certain terminology requiring some explanations in order for us to position it with respect to the points developed in the introduction: We use material interfaces of virtual reality, which we call **"behavioural interfaces"** because they exploit human behaviour. We prefer the "behavioural interface" term to "user interface" term. They are made of "sensorial interfaces", "motor interfaces" and "sensorimotor interfaces". In sensorial interfaces (visual interface, tactile feedback

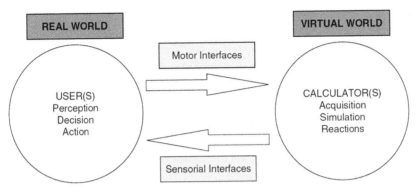

Figure 2.1 The "perception, decision, action" loop going through the virtual world.

interface, audio interface, etc.), the user is informed about the development of the virtual world through his senses. A visual interface is always used: VR headset, CAVE[1] or screen). Motor interfaces inform the computer about man's motor actions on the virtual world (joystick, data glove ...). Sensorimotor interfaces work in both directions (force feedback interface[2]). The number and choice of these interfaces depends on the objective of the application. **Real-time interaction** is achieved when the user does not perceive the time lag (or latency) between his action on the virtual environment and its sensorial response. This constraint is difficult to fulfil.

The user must be in the most effective "**pseudo-natural immersion**" possible in the virtual world. The immersion cannot be natural because we have learnt to act *naturally* in a *real* world and not in a virtual world (sensorimotor biases are created, that is why the term *pseudo*). This sensation is partly a subjective notion which depends on the application and the device used (interfaces, software programs, etc.) We will discuss at length the concepts of immersion and interaction which must be well defined and analysed at various levels.

We can derive a fundamental principle of virtual reality from this analysis. This principle is given in the loop shown in Figure 2.1. The user acts on the virtual environment by using the motor interfaces which capture his actions (gestures, movements, voice, etc.). These activities are transferred to the calculator, which interprets them like a request to modify the environment. In compliance with this request for modification, the calculator assesses the changes to be made to the virtual environment and the sensorial reactions (images, sound, effects, etc.) to be transferred to the sensory interfaces. This loop in interactive virtual environment is only a transposition of the "perception, decision, action" loop of man's behaviour in a real world. But two major

---

[1] CAVE: Cave Automatic Virtual Environment is an immersive virtual reality environment where projectors are directed to between three and six of the walls of a room-sized cube.
[2] A "force feedback interface" applies forces to user's hand who handles a virtual object. See the book "Virtual Reality: Concepts and Technologies", publisher CRC Press.

constraints, which are inherent to the techniques, disturb the "perception, decision, action" loop and consequently the subject's behaviour: **latency and sensorimotor discrepancies.** Latency is the time lag between the user's action on the motor interfaces and the perception of the consequences of this action on the virtual environment through sensorial interfaces. The existence of latency in the loop has an influence on the quality of any virtual reality application. This latency is an artefact inherent to interactive virtual environments. Sensorimotor discrepancies are the other artefacts of virtual reality. No matter how many sensory channels are used in an application, no matter how many interactions are provided to the subject, sensorimotor discrepancies with respect to the sensorimotor behaviour of the subject in the real world almost always exist. Do these sensorimotor discrepancies disturb the behaviour of the subject? These two issues are covered in chapter 8.

2.2 FUNDAMENTAL APPROACH FOR IMMERSION AND INTERACTION

2.2.1 The hierarchical 3-level model

After presenting the purpose and the technical definition of virtual reality, we will now explain our method for designing and assessing effective VR systems. Let's not forget that the purpose is *to help a person (or persons) to perform a sensorimotor and cognitive activity in an artificial world.* Thus it is necessary to first specify man's behaviour in a real world before going on to virtual world. Physically, man perceives his environment through his senses. A stimulus is received by a specific sensory organ (eyes, skin, nose, etc.). This sensory entity is the starting point in the transfer of information in the body through the nerve tracks. The afferent nerve message coming from the receiver is transferred to the nerve centres (spinal cord, brain). After integrating and processing the information, these centres transfer the efferent nerve message to the effector organs (skeletal muscles, ocular muscles, muscles of vocal cords, etc.). These organs perform the motor behaviour which results in movements. Man acts using his muscles and perceives the world through his senses, which are in fact more than five, if counting them is worthwhile! The kinesthetic sense, which is often ignored, creates problems in the static transport simulators and other VR devices. The sensory field of proprioception must not be ignored when we work on virtual reality. All sensory perceptions must be known and their impact must be studied for any VR application, even if not all of them are simulated. This often leads to sensory or sensorimotor discrepancies which should never be underestimated. The senses do not function independently of one another. Elsewhere[3], we explain in detail the concepts of behavioural interfacing, immersion and interaction in interactive virtual environments to deduce a model of man interfacing with these environments. We can use this model as a reference for designing and assessing any virtual reality application, with or without VR headset.

---

[3] "Virtual Reality: Concepts and Technologies", P. Fuchs, P. Guitton and G. Moreau, publisher CRC Press.

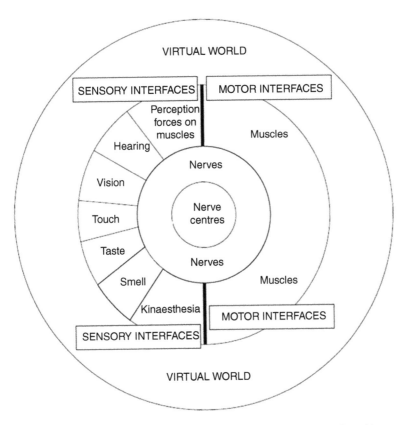

Figure 2.2 Anthropocentric diagram of man's vision of the virtual world.

Cognitive study of immersion and interaction in virtual environment is based on the approach of the subject's *activity*. Man is at the heart of the system since the virtual application is meant for him. It is the anthropocentric vision for the application user. In the absolute sense, it is possible to schematise (Figure 2.2) man as being completely immersed in an artificial world, the way he should perceive it **as a user.** But for the VR application **designer,** this anthropocentric vision is not enough. He must use this anthropocentric diagram, an objective to be achieved for the user, both by splitting and completing it because it is necessary to finely analyse the interfacing process and the devices to be designed. We will study behavioural interfacing using a technocentric diagram. The designer must keep switching between an anthropocentric approach and a technocentric approach to create an application in virtual reality. Knowing and mastering this duality of virtual reality helps to better understand the difficulties and the possible failures in making a human being become immersed and interact in an artificial world.

To have an in-depth understanding of this issue, we have made a fundamental choice of analysing this process at three levels. At first, we can schematise the interface between man and the virtual world *at the physical level*. In this case, we talk about

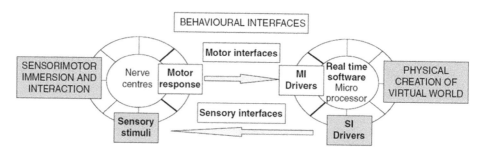

Figure 2.3 Technocentric diagram of sensorimotor immersion and interaction (Level 1).

sensorimotor immersion and interaction (Level 1). We obtain a "sensorimotor loop[4]" between the subject and the computer. Disruptions caused by latency and sensorimotor discrepancies should be as limited as possible. We obtain the **technocentric diagram** (Figure 2.3) at the physical level of sensorimotor immersion and interaction. We can observe that at the sensorimotor level, we disrupt the user's "perception, cognition, action" loop, by incorporating the artefacts (interfaces, their drivers and one or more computers). This diagram is restrictive as it gives only a partial representation of problems and solutions that every designer must study. This is a limited approach, which is seen to be used by certain designers who focus mainly on the technical aspects of virtual reality. If virtual reality exists thanks to technical developments, the technical problems should not be the only problems covered, ignoring the rest.

Behavioural interfacing poses a problem of interfacing similar to that of an operator with his machine or tool. In this case, thinking only about the physical interfacing (control buttons and feedback on actions) is not enough. It is equally necessary to understand the mental models according to which the person will think and act. In his **instrumental approach** of interfacing, Rabardel (1995) describes the instrument (interface) as a mediator of activity. In concrete terms, this activity is performed **physically** by effective motoricity and perception between man and the behavioural interfaces. These interfaces depend on artefacts (or instruments), and the user operates them using his cognitive processes. But which cognitive process would the immersed subject use in this situation? Will it be the cognitive process imagined and hoped for by the designer? Can the user master it and use it efficiently? This is one of the major difficulties of interfacing in virtual reality.

In principle, we would like to offer the user a relatively natural interaction and immersion so as to make them effective with respect to the application. In this case, we will talk about a **pseudo-natural** immersion (and also the interaction). What does this concept cover? First of all, we can confirm that whatever is done naturally in the real world is done unconsciously and without a great mental effort. We can, in principle, think that it will be the same in case of a virtual environment if the interaction and immersion are relatively similar to the human behaviour in the real world. Let's take an

---

[4]Generally, a number of sensory methods are used. Therefore, we should rather use the plural, "loops".

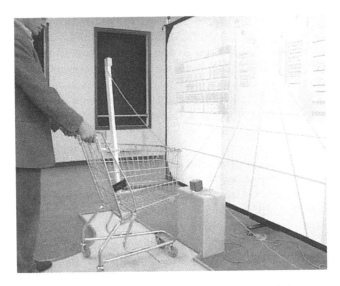

Figure 2.4 A real trolley is used to move in a virtual shop.

example: In a virtual shop, the consumer wants to go round the aisles. A real trolley is offered to him as a behavioural interface which he will push and take down the aisles, facing the screen. In this case, the consumer will unconsciously and naturally use some of the automatic reflexes he has acquired while shopping in real shops in this virtual shop with a few sensorimotor differences (Figure 2.4). That is why we use the prefix "pseudo" to refer to this type of immersion and interaction. The user thus uses a schema that he has adopted in the real world in his sensorimotor activity. The concept of schema is proposed by psychologist Piaget (Piaget & Chomsky, 1979). On analysing the origin of intelligence in an infant, mainly in its sensorimotor dimension, he states that the subject uses the schemas as the means to adapt to situations and objects he comes across. We would like to draw a parallel between the way a user understands a virtual world and the way a child understands our real world. For Piaget, a schema is a structured set of characteristics of an action that can be generalised, which helps to repeat the action or apply it to new contents (as in the case of a user who operates his trolley in conditions similar to the real world). It is on this concept that we base our approach to obtain behavioural interfaces, offering a pseudo-natural immersion and interactivity. Behavioural interface is thus a mixed entity including both an artefact (its hardware device) and a schema, which we call **"Imported Behavioural Schema"** (IBS). This schema is imported from the real environment to be transferred and adapted to the virtual environment. Piaget concludes that sensorimotor intelligence manages to solve a set of problems of actions (reaching an object, etc.) by constructing a complex system of assimilation schemas and to **organise the real world** as per a set of time-space and causal rules. Hence our fundamental postulate of virtual reality:

In an interactive virtual environment, a person uses the same approach that he uses in the real world to **organise the virtual world** *as per the set of time-space and causal rules.*

In case of technical, economic or theoretical difficulties that obstruct the use of an Imported Behavioural Schema, we can get around these difficulties by using a "**metaphor**". Instead of using a sensorimotor behaviour and the person's knowledge, we offer him a *symbolic image* of the action or of the desired perception visually in most cases. For example, in a virtual shop, we can give the consumer the opportunity to confirm the purchase of a product simply by clicking on its image and then on an icon representing a cash box. This action becomes symbolic and no longer represents the sensorimotor action in a real shop. Here, the immersion and interaction are less pseudo-natural. Using a metaphor may require more cognitive efforts if the metaphoric symbols are not known to the users. They need to make an effort to understand and assimilate the symbol, so that it gradually becomes a schema of use. But an Imported Behavioural Schema (IBS) itself can require certain efforts, as it must be adapted to a virtual world with a certain artefact and under the constraint of sensorimotor discrepancies. We use either a metaphor or an IBS depending on the psychological, technical and economic difficulties and the planned application. In practice, we can have a combined use of metaphors and IBS, as per the type of activities required (**cognitive immersion and interaction – Level 2**).

Another metaphor example: travel in the virtual environment is carried out via "teleportation" from one place to another in the virtual environment, while remaining motionless in the real environment. The place of arrival is generally selected using a localisation sensor with one hand, indicating the point of arrival virtually (metaphorical process). The place of arrival must be visible from the place of departure, otherwise you will to point to a map of the location. The user goes virtually from the place of departure to the place of arrival instantaneously.

At the third level, the objective is to attempt to *immerse the person in a given task (or a functionality)* and not a mere mental immersion in that virtual world. In this case, we talk about **functional immersion and interaction (Level 3)**. This division helps us to better clarify different problems faced in immersion and interaction of a subject. They are closely related and not opposite. To better understand their connection, imagine that you are grasping an object. We can use the schema of gripping (cognitive I^2),[5] using a six-degrees-of-freedom tracker and a screen displaying the object (sensorimotor I^2). But if the sensorimotor I^2 are not of good quality (long response time between action with the tracker and the feeling of movement on the screen), the schema of gripping cannot be used. On the other hand, if the interfaces do not use the schema of gripping, the cognitive I^2 will fail even with the interfaces functioning accurately.

The foundation of our approach is based on this hierarchical 3-level model and also on a diagonal division between the subject and the virtual world: Parallel to various levels of sensorimotor and cognitive I^2 for the person, we have two levels of software functioning for the virtual world. The computer should manage the software part in real time (real-time hub and drivers for hardware interfaces) symmetrically to the sensorimotor I^2, to **physically create the virtual world**. At the level of functional I^2 vis-à-vis the application and its objectives, we need to ask ourselves, "Which

---

[5]Hereinafter, we will use the abbreviation I^2 for immersion and interaction.

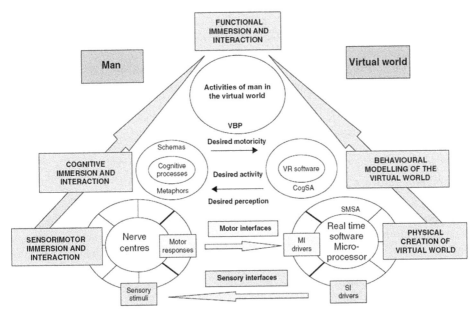

Figure 2.5 Technocentric reference diagram of VR with three I^2 levels.

activities should the user perform?" If you want to think over it for some time, you can quickly see that in all VR applications, the activities of the subject can always be divided into some basic behaviours that we call the "Virtual Behavioural Primitives" (VBPs). Therefore it is necessary to correctly define the VBPs and their characteristics at the functional I^2 level. They can be divided into four categories irrespective of the application:

- Observing the virtual world;
- Moving in the virtual world;
- Acting on the virtual world;
- Communicating with others or with the application.

The VR headsets are designed in order to improve the "Observing the virtual world" VBP with using a schema. But there is a difficulty: the VR headsets have a low field of view now (100° horizontal instead 180°!) (see chapter 7). "Moving and acting in the virtual world" with a VR headset must to be studied to avoid sensorimotor discrepancies (see chapter 8 and chapter 9). The "Communicating with others or with the application" VBP is not specifically studied in this book. Finally, we obtain the reference diagram of virtual reality (Figure 2.5) which we can use as a canvas in our design approach (Fuchs, 1999), with or without VR headset.

The CogSA and SMSA are explained in the following paragraph.

2.2.2 The behavioural software aids

To make the use of behavioural interfaces easier, one must not forget that the designer, via computer, can help the user to effectively use these devices. A number of possibilities can be programmed and used. For example, we can add specific constraints to the movements of a virtual object, operated by a six-degrees-of-freedom sensor, to help the user obtain the desired motoricity (the object's supporting surface will be parallel to the table surface when they are close to each other). All these interface using aids are grouped under the generic term **"Behavioural Software Aids"** (BSAs). Some of them will focus on a sensorimotor aid (SMSA) while the rest will focus on a cognitive aid (CogSA).

Sensorimotor software aids

The interface devices are not perfect in data transmission. Similarly, the sensorimotor behaviour of the subject assessed in the virtual environment is not perfect, or at least different from the one in the real world. The Sensorimotor Software Aids thus help to improve the interface, either by increasing the quality of data being transferred between the subject and the computer, or by proposing a "psychophysical" improvement in the sensory stimuli or motor responses. In the first case, the quality of a signal issued by the interface devices must be improved, for example:

- Screening the measurements of a location sensor is often necessary to delete the signal noises which could disturb the user while handing a virtual object. If the user blocks the movement of the object, the measurement noises should not move the object image even slightly. But data screening leads to increase in latency of the "perception, cognition, action" loop. Hence one has to try finding an optimum solution.

In the second case, the psychophysical improvement of the sensory stimuli or motor responses involves helping the subject, by using software support, to have a more comfortable sensorimotor behaviour in the virtual environment, for example:

- An adaptive spatial frequency filtering of stereoscopic images for their fusion by the brain (stereopsis) gives a 3D view to the subject, reducing the ocular stress (refer to chapter 9).

Cognitive software aids

The Cognitive Software Aids for motoricity and perception, associated with the VBPs, help the subject in completing a task. The CogSA can focus either on helping to solve the sensorimotor discrepancies or on helping the user to complete a task by detecting his intention. The following example illustrate the first case:

- For fixed transport simulators having sensorimotor discrepancies between vision and kinaesthesia, IFSTTAR (the French institute of sciences and technology for transport, development and network) recommends setting up a higher force feedback on the simulator's steering wheel compared to the feedback received in a real vehicle. In this way, the inexperienced drivers using the simulator have a greater perception of the vehicle's on-road behaviour though their vestibular systems have no knowledge about the vehicle movements. In this case, it is obvious that at the level of functional I^2 the force feedback does not need to be exactly like the feedback in the real world. The CogSA used helps every inexperienced driver using the simulator to control the vehicle.

In the second case, it is necessary to detect the intention of the subject who is trying to accomplish a task, for example:

- In a VR-based training system for the members of the driving crew of TGV, the user has to go near a telephone pole to pick up the receiver. Considering the interfaces used (big screen, moving walkway and handle bar), it is difficult for the driver to move as easily as he walks in a real environment (without BSAs, he might desperately go round the pole several times while controlling the handle bar). CogSA has been programmed to make this task easier: once the pole comes close and the driver's intention becomes clear, he automatically starts facing the pole. This is in line with the objective of functional I^2 because the training is not for moving in a virtual environment! Determining the intention of the immersed subject(s) is an important issue of research in the field of VR, which should eventually make it possible to find efficient cognitive software aids. The CogSA can be determined depending on the *affordances* of the objects or the subject's surroundings. As per Gibson's theory (1979), an affordance represents the interactions possible between the object and the subject. These interactions are perceived by the subject with respect to the idea that he has built about the object's function, more than what he perceives through the physical, geometrical or other characteristics of the object or the environment. It is thus more useful to know in advance the purpose of an element of the environment than having precise notions of its geometry or its physical characteristics.

Before using this model of interfacing of a user in a virtual environment for designing as well as assessing a VR device, it is necessary to remind the reader that the main difficulty in immersion and interaction of the subject comes from the disturbances in the "perception, cognition, action" loop(s). These disturbances can be overcome by correctly choosing the interface devices, cognitive processes and the BSAs to help the subject to act efficiently in the artificial "perception, cognition, action" loops that include an artefact (refer to figure 2.10). The interfacing of a user can imply comfort and health problems with the sensorimotor discrepancies. Chapter 8 is devoted to these problems, followed by chapter 9 that gives a detailed discussion of methods and 32 solutions to remove, or to decrease the VR sickness.

2.3 IMMERSION AND PRESENCE

Though we will not expand upon the concepts of presence and immersion here, a few key points[6] regarding these notions must be covered. While users make use of

---

[6]The notion of presence is at cognitive level I^2 in model $3I^2$.

a cognitive process to (inter)act via interfaces in VEs (cognitive level I^2), they also process information about the actions the application entails (functional level I^2). These cognitive aspects aside, users will nonetheless experience presence to a greater or lesser degree, especially in virtual environments far removed from those of everyday life. Does this sense of presence in a virtual environment cause feelings of pleasure or fear, or some altogether different emotion, a sense of being in a real but simulated elsewhere, for example?

Our experience of presence depends, in part, on notions of physical immersion in artificial space. Depending on the VR application, several degrees of physical immersion can be imagined:

- **presence, or visual immersion:** visual observation through a VR headset (without the user's hands represented), head-tracked motion alone creates interaction. This is the kind of presence felt in 360-degree video and VR applications simply meant to provide place or product observation, as in industrial project reviews. But such applications are generally seen in large screen environments and rarely via a VR headset;
- **presence, or semi-physical immersion:** users wearing VR headset see virtual representations of their hands, which can be used to grasp objects; it is a more immersive VE experience, some might argue. Semi-physical immersion is also possible without a VR headset. For example, observers can see their actual hands through a semi-transparent screen, placed directly opposite them at chest level, onto which virtual objects are projected in stereopsis. The users' hands are tracked to allow object manipulation.
- **presence, or full physical immersion:** a VR headset is worn to create a sense of presence and immersion. The user's body needs to be visually represented its entirety; it can make free use of all of its members. This type of immersion requires the user's hands and feet to be tracked. Here, too, VR headsets are not a prerequisite. For example, the user can be immersed in a CAVE, which allows the entire body to be present naturally. This is the CAVE's primary advantage over the VR headset, the second advantage being its higher image resolution, which is unmatched by current VR headsets[7].

But can we speak of "**total immersion**" simply because VR headsets are being used? To do so, a minimum of three conditions must be met:

- The representation of the subject's body in the VE must be appropriately rendered in real time;
- The sense of visual immersion must be total, requiring the field of view of the VR headset to match that of the immersed participant;

---

[7]While the visual quality of VR headsets will improve in the future, we can also hope that the cost of the CAVE will decrease when flat screens will be used instead of unwieldy video projectors.

–   The VE must spatially correspond to the RE to ensure the accessing of the subject's proprioceptive sensations (muscular and kinesthetic). Only thus can a sensory experience of a coherent environment be delivered, one merging the VE and the RE – the participant's body being physically present in the RE, of course!

But to better understand this last condition, we must take a closer look at our so-called five senses, of which, as mentioned earlier, there are more than five. This is the subject of the following chapter. Past and on-going studies have looked at the philosophical, psychological (Mestre & Vercher, 2011) and technical aspects of the sense of presence.

Chapter 3

Human senses

3.1 INTRODUCTION

We have already seen that a virtual reality application is designed on three levels of Immersion and Interaction, the first level being the sensorimotor I^2. The techniques of virtual reality use behavioural interfaces (sensory and/or motor) to physically join the computer with a human being. The design of such interfaces aims at creating a hardware device that has efficient metrological characteristics that must, if possible, correspond to the maximum psychophysiological capacities of senses and/or human motor responses. Under these conditions, it is indispensable to have a clear and precise image of the human sensorimotor behaviour in the real world to effectively analyse the behavioural interfaces as well as the sensorimotor I^2. Hence we are going to discuss the characteristics of specific sensory organs (vision, hearing, touch, smell and taste) and proprioceptive organs that allow spatial location, balance, movements and displacements.

A human being perceives his environment through his senses. Reception of a sensory stimulus is the starting point in the transfer of information in the body through the nerve tracks. The afferent nerve message coming from the sensory receiver is transferred to the nerve centres (spinal cord, brain and cerebellum). After integrating and processing the information, these centres transfer the efferent nerve message to the effector organs (skeletal muscles, ocular muscles, muscles of the vocal cords, etc.). These organs perform the motor behaviour, which results in movements, except in case of the muscles of the vocal cords that enable speech.

Every sensory impulse begins at specialised receptors. These receptors can be classified into two types. Some receptors, spread over our entire body, give information about the state of the body and its overall senses (somesthesia). Other receptors, which are specific to one sense, are located in the corresponding sensory organ. An example is the photoreceptors of the retina for vision. The exteroceptive receptors, which respond to external stimuli, inform us about our environment. Proprioceptive receptors respond to the actions on the body: body movements, its position in space and the forces exerted on the muscles. There are five specific sensory organs but more than five senses with all proprioceptive receptors. It is very important to know all human senses in order to design a VR application.

Researchers like Sherrington (Sherrington, 1906) suggest a strict distinction between exteroceptors, proprioceptors and interoceptors. Exteroceptors (eyes, ears, nose, mouth and skin) inform us about the changes in the surroundings and serve as the basis of perception. Proprioceptors (tips of organs, muscles, joints, internal ear) give us the sensations related to the position and movements of the body: they create the sensation of movement. Interoceptors (nerves leading to viscera) give vague sensations about the internal organs. Contrary to the position taken by Sherrington, Gibson (Gibson, 1966) suggests that sensing an action and sensing a movement do not depend on specialised receptors. The eyes, ears or the skin can note the behaviour of an individual as also the external events. For example, eyes note the movements of the head (front, back, rotation) through the movements of the surrounding light, optic flow (exteroception). Similarly, the movements of the joints or the internal ear can note the movements imposed on the body as well as the movements initiated by the individual. Hence, the proprioception – considered to be the means of obtaining information about our own movements – does not necessarily depend on proprioceptors. And exteroception – considered to be the means of obtaining information about external events – does not necessarily depend on exteroceptors.

There is a minimum level below which any stimulation does not create any effect on a sensory organ. Above this level, the minimum perceptible variation of a stimulus is proportional to the absolute value of the intensity of the stimulus (Weber's law). In terms of quality, this law means the following in the case of vision: For a light of low intensity perceived by the eye, a small variation in the intensity can be detected, whereas for a light of high intensity perceived by the eye, the variation needs to be of high intensity for it to be detected by the eye. In addition, a stimulus must last for a minimum duration for it to be perceived. On the other hand, if a stimulation is maintained constant, generally the sensation will disappear or will be reduced (adaptation phenomenon), except in the case of pain and certain nerve fibres.

Besides, when we discuss the resolution of user interfaces, we will see the importance of the density of receptors and its local variation in the sensory organs. The accuracy of human senses is equally important to be understood to prepare suitable sensory interfaces. In general, the absolute accuracy of senses is low in comparison with the relative accuracy since humans have a high capacity of comparing two stimuli, for example:

- It is difficult to define a colour of a single object, but it is easier to notice a slight difference between two colours observed simultaneously;
- Absolute perception of depth is difficult to estimate compared to the detection of a slight difference in depth between two neighbouring objects;
- It is difficult to quantify the absolute temperature of air or water coming in contact with our skin. However, we easily detect a slight difference in temperature between two fluids that are observed simultaneously.

The maximum frequency Fm of variation of a sensory stimulus perceptible by humans is an important characteristic to be understood. Whenever it is technically possible, the sensory interfaces should have a frequency band of 0 Hz to Fm. For example, it is recommended with a VR headset to display the images at the rate of

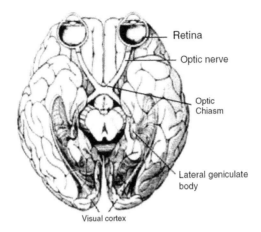

Retina

Optic nerve

Optic
Chiasm

Lateral geniculate
body

Visual cortex

Figure 3.1 Retinal projections towards the lateral geniculate body, then to the striate cortex.

more than 30 images per second (60 for stereoscopic images) which allows an animated
virtual world to be viewed in a continuous flow.

In this chapter, we do not explain all the knowledges about the senses[1], but only
some main characteristics in order to correctly use the VR headsets.

3.2 VISION

3.2.1 The human visual system

The two eyes, the sensory receptors of the visual system, participate in the observation
of space. The optic nerves from the eyes come together at the optic chiasm where the
left temporal fibres meet the right nasal fibres and vice versa. The fibres of the optic
nerve are then directed towards the lateral geniculate body. From the lateral geniculate
body, the information is sent to the occipital cortex. Figure 3.1 shows the path of visual
information from the eye to the visual cortex.

3.2.2 The eye

The function of the eyes is to channel the light of a wavelength in the range of 400
to 700 nm, emitted or reflected by an object to create a clear image which is printed
on that part of the eye which is covered by sensory receptors, i.e. the retina. The eye
is composed of a series of reflective media that play the role of a convex lens whose
total focal distance can vary with a modification of the curve of the crystalline lens
(Figure 3.2).

---

[1]See the book "Virtual Reality: Concepts and Technologies", publisher CRC Press.

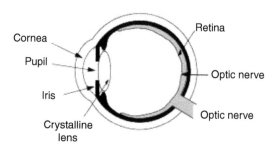

Figure 3.2 Anatomic structure of the human eye.

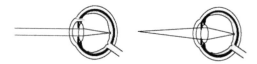

Figure 3.3 Accommodating infinity and accommodating a short distance (simplified optical diagram).

In a broad sense, the eye is a spherical darkroom. There are several dioptres at its entrance and the receptor structure at the rear:

- The pupil is the diaphragm of the system. The light rays are then projected on the retina which serves as a spherical screen;
- The cornea is a fibrous, transparent membrane which constitutes the main lens of the optic system. It is the anterior pole of the eye and the optical axis passes through its centre. Its horizontal diameter is 11.5 to 12 mm;
- The crystalline lens is a biconvex transparent lens placed between the iris and the vitreous body.

In order to explore space visually, an effectiveness of the eyes' movements is important. Several ocular movements make it possible to carry out these tasks: movements to modify the gaze direction and the movements of vergence to modify the depth of the vision. The movements to modify the gaze direction are carried out by saccadic eye movement which are the fastest movements of the human body (their speed can reach 500°/s).

3.2.3 Accommodation and vergence

While looking at an object less than approximately 100 meters away, the reflected image would become blurred on the retina if the eye did not have the ability to accommodate automatically. As a result of the action of the ciliary muscles, the power of the crystalline lens varies and makes it possible to focus on the retina to see objects that are near or at a distance (Figure 3.3): This phenomenon is called accommodation. The crystalline lens is the only dioptre of the visual chain whose power is variable. All light rays are diverted towards the fovea, which is the central point of the retina. The level of accommodation is adjusted to obtain a clear image.

The muscles of the orbital globes make it possible to orient the two eyes by verging them to the point in the space observed. This phenomenon is called vergence (or convergence).

As in accommodation, the vergence of eyes is done subconsciously, except if the person wants to squint. The movement of eyes in the orbital globes to change the vergence point and/or to follow a moving object can be very quick, with a maximum speed of approximately 500 degrees per second. Since accommodation and vergence of eyes is related to the depth of the object being observed, there is a relation between the two. This correspondence is not innate in humans, but acquired experimentally and subconsciously by infants. We are primarily concerned with the cross-coupling between accommodation and vergence. The accommodation imposes the vergence and reciprocally. When accommodation is stimulated alone by occluding one eye, the covered eye still converges via a coupling referred to as the accommodative vergence (AC/A ratio). This ratio is between 3 and 4. This natural relation can become artificially incorrect when we look at stereoscopic images on a screen: A virtual object placed behind the screen makes you verge the optical axes on it while the eyes adapt to the screen. This modification of the vergence – accommodation relation is likely to cause sensorimotor discrepancies and strain to the user. We will study this problem in paragraph 3.2.6.

3.2.4   The retina

The retina is the place where the light energy is transformed into discrete electrochemical signals. This signal comes out of the eye via the optic nerve. There are two types of photoreceptors – cones and rods; their distribution on the retina is very different:

- The cones, concentrated at the fovea, capture the wavelength. They are of three types, sensitive to wavelengths centred around 560, 530 and 420 nm respectively. Information of colour is produced at the cortical level by comparing the information received by various receptors containing different photopigments;
- The rods, on the other hand, are a lot less sensitive to colour. They are almost everywhere in the retina, but absent at the centre of the fovea and very dense at approximately 20° of visual angle.

Though the number of rods is much more than that of cones (120 million versus 6 million), the cones have a major contribution in the information transmitted to the deeper structures of the visual system. It is thanks to these cones and rods that this system can adapt to the ambient light intensity and cover a range of intensities close to 7 logarithmic units. Only the rods can operate when the illumination is very low. In such conditions, the cones are positively coupled to their neighbours to increase the surface of the photon detector. Sensitivity is increased at the expense of the resolution of the system. On the other hand, when the light intensity increases, the photoreceptors are decoupled so as to obtain maximum resolution.

The retina has several layers of neurons that cover the bottom of the eye. These layers form a carpet of cells connected horizontally within the same layer and vertically from one layer to other creating a strong pyramid-shaped neuronal architecture. In fact, if the number of photoreceptors is in the region of 100 million, the number of ganglion cells whose axons constitute the optic nerve is approximately a million. All axons of the ganglion cells come out of the retina in the same region that has no photoreceptors, called the blind spot or the optic disk.

3.2.5 Vection and illusions of self-motion

The term vection is used most often to describe visual illusions of self-motion in physically stationary observers. Circular vection is typically used to describe visual illusions of self-rotation, whereas linear vection is used to describe visual illusions of self-translation. The visual system registers the optical flow produced by the world moving past our head. When we move through real environments, we have no problem with maintaining awareness of our position despite the fact that our relationship with surrounding objects constantly changes. But when a large part of the visual field moves, a motionless man feels like he has moved. For example, when one is in a train at a station, and a nearby train moves, one can have the illusion that one's own train has moved in the opposite direction (see chapter 8).

3.2.6 Visual perception of depth

3.2.6.1 Cognitive perception by monocular cues

It is necessary to first understand that the perception of depth of a three-dimensional world is not only because of binocular vision. With only one eye, a human being can interpret the image received and deduce the notions of depth. This interpretation is done subconsciously through a cognitive processing thanks to what he has learnt from a very early age. At the early stage of visual processing, the decomposition and processing of visual information by different channels makes it possible to use a varied series of cues for the perception of three-dimensional space, and thus that of depth. The monoscopic films allow the perception of three-dimensional space and not only the stereoscopic films, called "3D" films. Although the term "3D" is ubiquitously used.

These cues can be divided into two main categories:

- Proprioceptive cues (caused by actions of orbital and ciliary muscles), made of **accommodation** and **convergence**: these cues are weak at a short distance (a few meters);
- Visual cues, made of **binocular and monocular cues.**

Proprioceptive cues, accommodation and convergence, are adjusted by the visual system. It is through the value of command of the orbital and ciliary muscles that the brain has a proprioceptive perception of the depth of the object observed. Let's not forget that the natural vision of human beings is an **active** vision. The eyes are almost always moving to observe the environment and the crystalline lenses change their form to facilitate accommodation. On the other hand, a camera observes the real environment in a **passive** vision. The depth of field of any photograph is a depth index that corresponds partially to the accommodation phenomenon. We can thus use these effects of depth of field in computer-generated images to give a more "3D effect" to the space observed.

Figure 3.4 Variations of light and shadows increase the "3D effect" of the two cubes.

Figure 3.5 Do you see 4 roses of same size more or less separated or 4 roses on the same plane?

In monocular vision, the monocular cues that are learnt subconsciously make it possible to understand the third dimension with one eye, even if the perception of depth is clearly better quantified with binocular vision. Monocular cues can be categorized in the following manner:

1 Light and shadows

The variations in light and shadows on the objects help achieve a better perception of the three-dimensional form of these objects (Figure 3.4).

2 Relative dimensions

Each object sends its image to the eye. The dimensions of this image are proportional to that of the object and decrease as per the distance with respect to the eye. The brain knows the "normal" dimensions of real objects. It can thus understand their approximate distances. Similarly, estimate of depth is facilitated for a series of real or imaginary objects of identical shapes: they are *basically* perceived at the same size and placed at different distances in a 3D space (Figure 3.5).

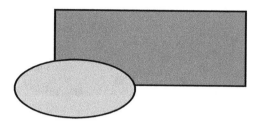

Figure 3.6   Occultation: you see an ellipse in front of a rectangle.

Figure 3.7   Variation of visibility and perception of texture.

3 Occultation or interposition

An object can partially hide another object placed behind it, which makes it possible to relatively position these objects in depth (Figure 3.6). The brain interprets this image through cognitive reasoning: it spontaneously perceives an ellipse in front of a rectangle and not as two adjacent shapes on the same plane.

4 Gradient of the texture of a surface

The texture of a surface is perceived clearly if the surface is placed at a slight distance. This gradient of the texture gives additional information about depth.

5 Variation of visibility in an outdoor scene

The visibility of an outdoor scene decreases with the thickness of the atmospheric layer, and hence the depth (Figure 3.7).

6 Parallax caused by movements

When an observer moves, the retinal images of stationary objects have relative movements depending on their distances. It is the same when the objects move relatively to each other. In practice, we can use this rule to show the depth by rotating a product or

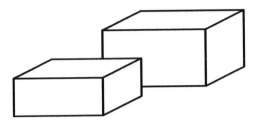

Figure 3.8 Perspectives.

an entire 3D scene shown on a monoscopic screen. We can also modify the images on the screen, monoscopic or stereoscopic, depending on the movement of the observer in front of the screen. But this solution is more difficult in practice because it requires detecting the position of the observer's head in real time. On the other hand, it is one of the interests to exploit a VR headset which allows in real time the modification of point of view and thus a better perception of depth. The reader has to note that this monocular cue is more powerful to perceive the depth than stereoscopic cue (see our work (Leroy, 2009)).

7 Perspective

This is certainly the most effective rule allowing the perception of a world in relief on a monoscopic screen. This technique is used in painting, right from the Renaissance era, to show three-dimensional space on a plane surface. It is worth noting that when the video game designers speak of "3D" video games, they change the plane images (without perspective) to images with perspective on monoscopic screens. There are various types of perspectives (isometric, geometric, photographic, artistic ...), see Figure 3.8.

Note: The reader can note that we have just used these rules which are shown in the graph in Figure 3.9. You can close one eye but the perception of depth remains the same! These rules are now frequently used in computer-generated images to give a three-dimensional representation on a computer screen. For example, in Open GL you can calculate your images in isometric perspective (orthographic projection) using gluOrtho2D command or in photographic perspective3 (perspective projection) using gluPerspective command.

3.2.6.2 Convergence and retinal disparity

Let's see the phenomenon of convergence and its effects on the retinal images: By staring at a finger held at 20 cm, you see a ghost image of any object in the background. On the other hand, if you stare at an object placed behind your finger, your finger looks split into two: the brain cannot merge two different images of your finger as the retinal disparity is too large (Figure 3.10).

As the visual fields of the two eyes (partially) overlap and the optical axes converge, we have two slightly different views of the same scene which helps us perceive the depth. When the optical axes converge on an object, the brain obtains ghost images of other

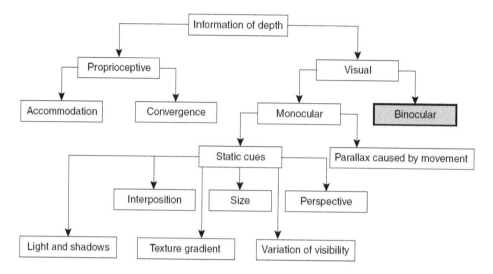

Figure 3.9 Cues that help in perceiving depth in the three-dimensional space.

Figure 3.10 Fixing on the finger or fixing on the tree.

objects distant from this object. What does the brain do with these ghost images? In practice, it takes no mental note of these images. Mainly, even when you look into the distance, you always see two blurred images of your nose but you do not pay attention to these images ... except now ... it will disturb you if you did so! In case of natural vision, the eyes do not accommodate ghost images (the corresponding objects are too far or too close). But it is not the same in case of a screen where all images are clear and are at a same distance. We will discuss this problem when we talk about creating 3D images. Normally, the convergence of visual axes is ordered by the brain,

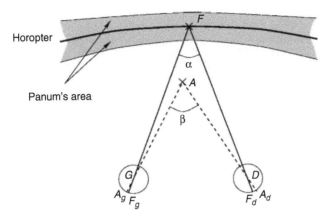

Figure 3.11 Perception of the difference in depth between fixation point *F* and point *A* on the basis of disparity.

but we can physically make these axes diverge and see that images get split: Look at this text and press your finger on one eye to turn it. The images will be split; merging them will no longer be possible.

Let's analyse our natural vision more geometrically and optically using Figure 3.11. Here, an observer looks at point *F*. The (horizontal) retinal disparity quantifies the notion of difference between images by measuring the distances. Figure 3.11 illustrates retinal disparity: for fixation point *F* that requires a convergence angle α, disparity d at point *A* subtending an angle β is defined by: $d = \beta - \alpha$, which corresponds to angle DF_dA_d when *A* and *F* are perpendicular to the straight line connecting *D* and *G*. In geometrical terms, the disparities are different enough only if the objects observed are close to the eyes. The binocular perception of depth is thus effective only at small distances (a few meters).

3.2.6.3 *Binocular vision and diplopia*

For a given fixation point, all points in the space projecting in the corresponding retinal positions form a horopter. The theoretical horopter is called the Vieth-Muller circle: it is the circle which passes through the fixation point and the nodal point of each eye. Determination of the empiric horopter depends on different criteria as per the definition selected. The points situated ahead of the horopter are said to be converging or in crossed disparity, whereas those behind the horopter are diverging or in direct disparity.

If the retinal images are near and have slight horizontal disparities, the binocular stimuli are perceived in a single image due to fusion of points that are within the Panum's area (area in grey in the previous figure). This is known as binocular vision. The objects are perceived to be in depth in front of or behind the fixation point by binocular vision. However, binocular vision and appearance of depth are damaged by vertical disparity. Beyond Panum's area, the stimuli of each eye are perceived separately. They are split, known as diplopia.

3.2.6.4 Neurophysiological mechanisms of the perception of depth

Depth can be estimated after mapping. This problem is far from being trivial because disparity is an ambiguous cue. In fact, the same amount of disparity can be associated to different distances depending on the convergence of vision. Therefore, to estimate the actual depth, the information of disparity must be graded as per the convergence. This process is called constancy of depth. Activity of cortical cells processing the disparity should thus be extremely specialised and modulated by the fixation distance. Trotter observes that when we modify the real distance of depth, the activity of neurons changes remarkably. Neurons always prefer the same retinal disparity, for example "nearer" or "farther" from the fixation point. However, their level of activity depends to a great extent on the distance from the object. Trotter shows that coherent perception of the three-dimensional space is the result of a combination of messages issued from the retina and the extra-retinal information regarding the position of eyes, made in a group of neurons in the primary visual cortex. The most probable information is the proprioceptive signal from the extrinsic eye muscles (Trotter, 1995).

To conclude this section on visual perception of depth, we can say the following regarding the phenomenon of "3D" vision: from a real 3D object, an observer receives two 2D images on his two retinas. His visual system deduces a 3D perception from these images. In an on-screen artificial 3D vision, we display two 2D images calculated on the basis of a computer-generated 3D model of the object. The retinas receive two 2D images which are perceived by the visual system as 3D!

3.2.7 Psychophysical characteristics of vision

In this paragraph we will specify some characteristics useful to understand a visual interface. These characteristics correspond to the efforts currently undertaken to improve visual interfaces (VR headsets and other devices):

- Improvement in the image definition;
- Increase in the visual field;
- Stereoscopic vision and
- Immersion of eyes.

3.2.7.1 Visual acuity

Visual acuity is not homogenous in the entire visual field. For an emmetropic eye (normal), monocular acuity is very high for a cone centred on a 2° angle since the distribution of retinal cones is restricted to the fovea of the eye. The minimum value of the angle from which the two points are seen separately depends on the stimulus observed:

- A strip of light on a black background: 30″ angle;
- Two points of light on a black background: 1′ angle;
- Two dark points on a light background: 2′ angle;

As we can see, the average value (1′) is in accordance with the density of photoreceptors in the fovea. An angle of 1 minute corresponds to the vision of two points separated by 0.1 mm for a distance of 35 cm between the image and the eyes. To follow these characteristics, a 25 cm wide screen positioned at this distance should display 2500 pixels horizontally. The actual computer screen has a good-quality of the image definition now. These numeric values are established for a screen giving a narrow visual field (40 degrees). But the image definition is insufficient taking into consideration the human visual system for VR headsets with the fields of view between 100° and 210°. If we want to increase the field while maintaining a sufficient screen resolution, we will have to display a huge number of pixels which is not compatible with current technology. The average value (1 minute of angle) of visual acuity must be referred to in order to judge the resolution quality of a screen or VR headset used in a VR device (see chapter 7).

Human binocular vision has the capacity to detect a difference ΔP in depth between two planes P and P'. Stereoscopic acuity has been defined:

$$\Delta P = 0.001 P^2$$

This means that if an object is located at 10 m from the observer, he can see the relief only for two planes at 0.1 m. For an object at 1 m, the relief can be perceived only for the planes separated by 1 mm. To conclude, three dimensional vision is relatively efficient only for short distances, for example when the user looks at his hands. Since this vision causes eyestrain, these considerations should force the designers not to *automatically* use stereoscopic vision in any given situation, but only when it is justified to use it. We will discuss this once again, but the reader should remember that stereoscopic vision, if not used correctly, can go against the set goal: better perception of depth for an effective visual immersion.

3.2.7.2 *Visual field*

Both the eyes take part in observing space. The area of space seen by either eye at a single instant is defined as the visual field. The points of space that are in the zone of binocular overlap are seen simultaneously by the left and the right eye.

The characteristics of the visual field for a stationary eye are approximately:

- 90° (temple side) and 50° (nose side) horizontally;
- 45° (up) and 70° (down) vertically;

The visual field for the two stationary eyes: 180°.

If we want to achieve visual immersion with a VR headset, it is necessary to take into account the movements of the eyes and the head. An eye can turn in its orbit by approximately 15° horizontally and vertically. For a full visual immersion, the headset has to a horizontal field of vision of 210° (see the "Star VR" headset, chapter 7). In practice, the field of vision in the majority of VR headsets is much smaller (about 100° horizontally). The impression of visual immersion is thus more or less strong with these devices. An eye can turn in its orbit at a maximum speed of about 600°/s. The head can turn at a maximum speed of about 800°/s.

3.2.7.3 *The Interpupillary Distance*

Remember that the Interpupillary Distance (IPD) is variable in humans. On average, it is 65 mm in adult men and a little less in adult women; the differences can be significant, between around 50 and 70 mm, in Europeans. This size should be taken into account while using the head-mounted displays, as the two optics must be correctly centred with respect to the eyes and thus have an optimum field of vision. For fine settings of a head-mounted display, we have measured the IPD for each user using a pupillometer. These measurements make it possible to correctly set the IPD, rather than letting the user make approximate adjustments.

3.2.7.4 *Power of accommodation*

Accommodation is the process which the changes optical power to maintain a clear image or focus on an object as its distance varies. Distances vary for individuals from the far point—the maximum distance from the eye at which a clear image of an object can be seen, to the *near point*—the minimum distance from the eye at which a clear image of an object can be seen. The young human eye can change focus from distance (infinity) to as near as about 7 cm from the eye. The amplitude of accommodation declines with age (25 cm in about 40 years). Thus the VR headsets have optical devices with a large visual field for the accommodation of eyes on screens that are very close to them. With respect to these two requirements, the technical solutions are often opposed to each other: It is difficult to obtain a large visual field with mini-screens that are very close to the eyes.

3.2.7.5 *Maximum temporal frequency in vision*

Temporal frequency of images to perceive a flow of movements is critical for values below 25–30 images per second (a value depending on the type of images). The movements will be free-flowing on monoscopic screens at 50–60 Hz (an image is made of two frames). In case of stereoscopic screens with time-division multiplexing, the scanning rate must be double: 100 Hz. The frequency of images displayed (FPS: *Frames per second*) must not be too low compared to the needs of the visual system to perceive images without flickering and with continuously moving objects in motion. This is not dependent on retinal remanence (old discounted theory) but on the "phi" phenomenon and the "beta" movement, which are neurophysiologic mechanisms: when the brain perceives, for example, two bright spots alternately displayed at an angular distance close to one another, the brain perceives the first point moving to the position of the second point, which gives an impression of movement on the basis of one-off information. The higher the frequency (FPS of 120, 240 Hz), the more movements will appear fluid.

3.2.8 Psychophysical characteristics of stereoscopic vision

As explained earlier, the visual system can merge the two retinal images for all the points located in the Panum's area. In other cases, the brain either cannot merge the two images, or only does so with a certain amount of strain. For a three-dimensional vision, the aim is to create stereoscopic images by creating small retinal disparities in

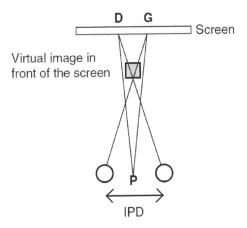

Figure 3.12 Horizontal parallax is defined by the DPG angle.

the observer's eyes while maintaining an effect of depth. What are the limits of merging generally acceptable for human beings?

We have previously defined retinal disparity. For 3D images displayed on a stereoscopic screen, the horizontal disparity is (approximately) equal to the horizontal parallax. It is defined by the DPG angle, formed by the two homologous points of left and right images, seen from the viewpoint of the observer (Figure 3.12).

Yei-Yu Yeh and Silverstein (Yeh, 1990) included the criterion of the duration of viewing the images to examine the limit of merging and the estimation of depth. The limit of merging is analysed on the basis of different parameters of stimuli displayed on a cathode screen with active eyeglasses: colour, display time and horizontal parallax. The results imply that the limit of merging (given in angle of parallax) is very low for brief stimulus of 0.2 second, i.e. 27 to 24 minutes of arc compared to a stimulus of 2 seconds, i.e. 1.5 degrees. It is therefore more difficult to observe rapidly moving 3D images. Taking the horizontal parallax as the parameter influencing the binocular vision, experimental studies (Valyus, 1962) have demonstrated the difficulty in merging two plane images having horizontal parallaxes (and hence retinal disparities) of an angle higher than 1.5 degrees. We carried out tests in working conditions as part of the remote-control operations with real stereoscopic images. Our results give 1.2° (Fuchs *et al.*, 1995) as the limit for horizontal parallax, a value that is a little below the limit defined by Valyus. These variations are normal as they vary from person to person, depending on their tolerance for 3D images and on whether they force their visual system to merge the images. It should be noted that 1 to 5% (possibly more) of the population cannot merge 3D images on a screen. The studies show that it is difficult to merge images creating vertical parallaxes (and thus vertical disparities). Eyes, generally placed on the same horizontal plane, are not capable of perceiving vertical disparities, except if they are very low. For 3D images, vertical parallaxes must be less than an angle of 20′ arc (Julesz, 1971). We will see that this constraint needs to be considered when we talk about creating 3D images.

The results of the studies conducted at l'IRBA (Institut de Recherches Biomédicales des Armées) of the French Defence Ministry, being more accurate, corroborate the previous results. The limit of merging also depends on the *spatial frequency*[2] *of images*, which is not studied in the previously mentioned works. On the basis of their studies, we can conclude that greater horizontal disparities can be merged when the visual simulations have low spatial frequencies. From a practical point of view, we can say that the objects represented by their outlines (high spatial frequencies) are more difficult to merge. Their studies have highlighted two mechanisms involved in merging, differentiated by the duration of stimulus: immediate merging for smaller disparities and non-immediate merging for greater disparities, putting the reflex vergence of eyes into play. The results also show that at a spatial frequency of 4.8 cycles/degree, there is immediate merging up to 20' of arc and maximum merging at about 52' of arc. At a spatial frequency of 0.22 cycles/degree, there is immediate merging up to 80' of arc and maximum merging at about 176' of arc (Perrin, 1998).

3.3 CUTANEOUS SENSITIVITY

3.3.1 The skin

Touch receptors are constituents of the skin. It will be particularly interesting to study the skin in detail because it is on the skin that one will have to work in order to reproduce the desired sensations: touch sensitivity, pressure sensitivity, vibration sensitivity and temperature sensitivity.

3.3.2 Classification of biological sensors

A criterion to classify cutaneous receptors is the reaction to a preferential stimulus. This classification creates multiple types of receptors of which three are known as nociceptors, thermoreceptors and mechanoreceptors.

Nociceptors

Nociceptors take care of the sensations of pain and are thus the safety devices indispensable for the protection of the manipulator system. It is difficult to differentiate between this category of receptors and the other touch receptors.

Thermoreceptors

Temperature is a parameter that evidently comes into play during tactile exploration. Man is a homeothermic animal; he carries out thermal exchanges through conduction, convection and evaporation. The sensations of heat or cold appear when we deviate from thermal neutrality. Thermal sensitivity is due to the thermoreceptors in the skin. They can be divided into two groups – cold receptors and heat receptors, spatially separated on the skin surface (expanse) and in its layers.

---

[2]Spatial frequency is the number of cycles of a periodic oscillation of light or colour in one degree of visual angle. Its value is given in terms of cycles per degree (cpd).

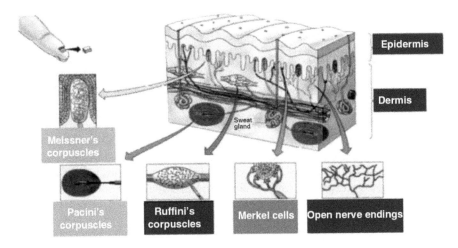

Figure 3.13 The skin with the mechanoreceptors.

Mechanoreceptors

The mechanoreceptors play an important role in tactile perception. These receptors are the ones that are most frequently called into action when a contact is established between the skin and an object (or material) from the outside world. These neurons are located in the entire body and can be free nerve endings or can also be closed in the form of small bulbar, axial or cylindrical corpuscles. The distribution of these mechanoreceptors and the combination of their efferent signals in response to the stimuli involve a specialisation of the sensory structure.

3.3.3 Density of distribution

The skin has a very rich afferent innervation; its density varies significantly from one place to other. For example, the tips and the flesh of our fingers have a large number of nerves (about 2500 receptors/cm^2). The surface of our hand that is used to grip is estimated to have about 17000 mechanoreceptors. The distance between them varies between 0.7 mm on the fingers and 2 mm on the palm.

3.3.4 Classification of mechanoreceptors

Mechanoreceptors in the human body are characterised by their functional aspect, their layout (distribution, location, etc.) or even by the nature of their temporal responses. Four types of exteroceptive mechanoreceptors have been identified (Figure 3.13). The first two are the Meissner endings located in our hand's skin between the papillary ridges and the dermis and the Merkel endings located at the end of these ridges. These two types are characterised by a large spatial resolution. The remaining two types are

Pacini's corpuscles and Ruffini endings. They are found deeper in the skin (i.e. below the dermis) with larger reception fields[3].

3.4 PROPRIOCEPTION

The three fields related to proprioception are the sensations of position in the space, body movement and forces exerted on the muscles. The first two sensations correspond to the kinaesthetic sense and the third sensation corresponds to muscular proprioception. The organs contributing to these sensations, in addition to the proprioceptive organs of the muscles, tendons and joints are the organs located in the vestibules of the internal ears. This is why we will talk about the anatomy and the physiology of the vestibular apparatus in this section. We will then discuss the problem of perceptive localisation.

3.4.1 Vestibular system

The vestibular system of the internal ears system includes two components:

- three orthogonal semicircular canals, which indicate rotational movements;
- and two otoliths organs on each side (included in saccule and utricle) which indicate linear movement.

The vestibular system sends signals to control eye movements (for example: the vestibulo-ocular reflex) and to the muscles that keep humans vertical. The Vestibulo-Ocular Reflex (VOR) is a reflex eye movement that stabilizes images on the retina during head movement by producing an eye movement in the direction opposite to head movement. Thus it conserves the images at the center of the field of view.

If the head is moving, the nerve message provides information about the total acceleration given to the head, which is the total of the acceleration of the movement and earth's gravity. This means that by inclining a person in a simulator cabin, we can indirectly simulate an acceleration of movement by visually showing him a virtual movement with a VR headset or with a large screen. In case of linear accelerations (in translation motion), the macular receptors (utricle and saccule) perform the discharge function. The semicircular canals are sensitive to angular accelerations (rotation). Alain Bertoz (Bertoz, 2000) explains that the coordination between the body position in the space and the body movements is complex. The stabilization of the head is essential thanks to vestibular system.

3.4.2 Articular proprioception

Other proprioceptive receptors, distributed all over the body, are sensitive to the position or movement of different body parts. The neuromuscular spindles and Golgi tendon organs in the muscles act as receptors. The receptors in the joints respond to

---

[3]For more information, see the book "Virtual Reality: Concepts and Technologies", publisher CRC Press.

variations in the stretch and are thus sensitive to relative movement between two body parts. The brain analyses the data coming from all the proprioceptive receptors with the sensory data sent by the skin, eyes and the internal ear. This is how the brain understands the body's position and its movements.

3.4.3 Muscular proprioception

The muscular proprioception is the sensitivity to the forces exerted on the muscles which gives an overall information about the forces of contact between the person and an object. The perception of the vertical direction is very important to understand the spatial body schema and thus to be able to assess the world surrounding us. Thus the movement perception is multimodal. If in a VR application (with or without VR headset) the proprioceptive organs are incorrectly stimulated, there are sensorimotor discrepancies. The muscular proprioception is the sensitivity to the forces exerted on the muscles which gives an overall information about the forces of contact between the person and an object. It is possible to use the force feedback interfaces (see Fuchs *et al.*, 2011).

We know that the brain anticipates the future sensory stimuli it receives, see (Berthoz, 2008). When users control their movements, the brain is able to anticipate. The brain has mechanical internal models. However, when users are passive relative to their movements and trajectory in virtual reality, it is difficult for the user to anticipate. It is a good idea to provide sensory cues to help anticipate any changes in acceleration (see chapter 9).

To give the illusion of being immersed, the presentation of the virtual world surrounding the subject should be subject-centred. He then perceives himself to be at the centre of this world. For this purpose, it is important to recreate the exact position of that subject's body in the virtual world. It is therefore necessary to use the sensors that transfer the data regarding the position of the body and legs in the real world to the computer controlling the virtual world. Similarly, considering the human sensitivity to movements in three directions, it can be necessary to restore to the subject immersed the sensations related to the simulated equilibration (vertical perception) and movements. It is possible to use the platforms for dynamic simulation of movements (see chapter 6).

In conclusion, it is important to understand the human senses in order to design an efficiency VR application without sensorimotor discrepancies (see chapter 8 and chapter 9).

Chapter 4

Visual interfaces

4.1 INTRODUCTION

Using the visual sense is almost always indispensable in virtual reality. Virtual reality software makes it possible to create computer-generated images of increasing quality. In terms of computer hardware, sufficient computing power is available to create good quality three-dimensional images in real time. The latest advances have been made in graphic cards, which currently help in real time computing of images that come close to photo-like rendering, though some technical hitches still remain to be overcome.

Before the following "VR headset" chapter, this chapter presents briefly the other visual interfaces. This enables to any designer of VR application a comparison between the VR headsets and the other types of visual interfaces to know if it is quite judicious to exploit a VR headset. We notice that some designers use VR headset without having studied the other possibilities of visual interfaces. There is no ideal and universal visual interface!

An ideal and universal visual interface must possess metrological characteristics that correspond to the maximum capacities of the human visual system so as to use this sensory channel effectively. We, however, have a very long way to go to reach this goal. This visual interface should thus offer four additional capacities over an ordinary screen:

- large horizontal and vertical fields of vision corresponding to those of our eyes;
- stereoscopic vision in the entire binocular field of vision;
- high graphic resolution using all the performances of monoscopic and stereoscopic acuities;
- and a gaze immersion[1] in the virtual world.

Achieving this is possible only if the visual interface is a VR headset which is connected to a sensor locating the orientation of the operator's head. The computer must obviously be sufficiently powerful to show him, in real time, left and right images of high resolution as per the direction of his eyes.

---

[1]Gaze immersion: The user always sees the virtual scene, even if he turns his head and his eyes in any direction.

4.2   PROCESSES FOR THE VISUAL PERCEPTION OF THE 3D SPACE

Before turning our attention to actual visual interface technologies, it is necessary to look at the processes involved in three-dimensional display. These practicalities are naturally related to the characteristics of human vision, including both monocular cues, allowing us to comprehend the third dimension with just one eye, and binocular cues, providing a noticeably better perception of depth (see Chapter 3). As mentioned previously, it is important to keep in mind that it is the *observer*, and not the technology, who provides the cues involved in three-dimensional spatial perception. To better understand the process of 3D perception and immersion, a certain number *display functions* (Figure 4.1) need to be reviewed:

- Function 1, "**monoscopic display**": Monocular cues, especially perspective, are used to project a three-dimensional space onto a flat surface, or monoscopic screen. This is the simplest way to achieve a three-dimensional display, but field of view is limited and gaze angle usually controlled by a handheld device. Most video games involve this function (Figure 4.1 – Number 1), namely a computer or game console connected to a joystick.
- Function 2, "**monoscopic display with gaze immersion**": A three-dimensional space is projected onto a monoscopic screen, but, in this case, the image is controlled by user movement of the head. Here, too, field of view is limited. The technology, which is not widespread though utilised for certain smartphones features, involves the monocular "motion parallax" cue to augment three-dimensional perception of space (Figure 4.1 – Number 2). Two combinations are possible:

  o The screen is stationary and user head position is tracked: User movement determines the on-screen angle of view;
  o The screen is mobile (smartphones and digital tablets) and is manipulated by the user: A position sensor measures screen location and modifies angle of view.

- Function 3, "**stereoscopic display**": Monocular and binocular cues are utilised simultaneously. This technique is typical of 3D film and 3DTV. Field of view is limited. As noted previously (see Chapter 3), the "motion parallax" cue provides a better perception of forms and distances in space, than the stereoscopic cue does. Function 2 is therefore more effective than Function 3 (Figure 4.1 – Number 3);
- Function 4, "**stereoscopic display with gaze immersion**": Techniques involved in Functions 2 and 3 are coupled with a screen to provide better perception of depth. Here, too, the field of view is limited (Figure 4.1 – Number 4);
- Function 5, "**passive immersion display**": Images cover the user's entire field of view. The angle of view is not controlled by the user. This technique is used for installations like the "IMAX Dome", with or without stereoscopic imagery (Figure 4.1 – Number 5);
- Function 6, "**display with active immersion**": Stereoscopic images cover the user's entire field of view; head-tracked motion controls the angle of view. VR headsets and the CAVE utilise this function. In practice, however, the field of view of the

N° 1 «*monoscopic display*»

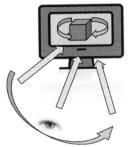

N° 2 « *monoscopic display with gaze immersion*»

N° 3 «*stereoscopic display*»

N° 4 « *stereoscopic display with gaze immersion*»

N° 5 «*display with passive immersion*»

Figure 4.1 The five first functions.

majority of VR headsets is much smaller that of human beings (respectively 100-degree and 210-degree horizontal fields of view). It is also worth noting that CAVE installations require users to wear glasses, which also interfere with user field of view. In short, current visual immersion interfaces are far from ideal.

Of these 6 functions, Function 6 is the one that most closely matches human visual perception. In the future an even closer 7th function is likely, because there will be less discrepancy between the accommodation cues involved in both natural and unnatural viewing processes. In functions 1 through 6, accommodation is unnaturally static, given that the user's gaze is focused on the surface of the screen, which provides inaccurate cues of distance and depth. Next generation "Light Field" screens (see the next chapter), enabling variable accommodation, will allow users to accurately determine the distance of objects observed in virtual environments. "Light Field" screens are being designed for a number of new AR[2] headsets.

By integrating eye-tracking devices into future VR headsets, it will be possible to process stereoscopic images and therefore improve the user's visual comfort and perception of the depth.

This description of the many visual functions involved in 3D perception demonstrates the complex nature of selecting a visual interface. It is not just a question of simply choosing between a standard screen and a VR headset.

The presentation of visual interfaces can be based on various classifications, depending on which of the four capacities of an ideal interface mentioned above is considered more important. For the first two classes of interfaces, we will take the interfaces that facilitate or do not facilitate immersion of eyes, i.e. VR headsets and fixed-support interfaces. The sub-categories depend on the vision fields offered by the interfaces and then on the possibilities of stereoscopic vision. The VR headsets are presented in the following chapter.

4.3 VISUAL INTERFACES WITH FIXED SUPPORT

4.3.1 Monoscopic or stereoscopic computer screens

A flat screen of a standard computer, which is monoscopic or stereoscopic, can be used for different domains closely related to virtual reality: Video games, etc. It should be noted that some screens have sufficient resolution in comparison to the visual acuity of humans, according to the field of vision used, which depends on the distance between the observer and the screen.

Irrespective of the technique used in stereoscopic vision, the objective remains the same: providing a different image for each eye. For this purpose, there are two principles: we can either put two small screens near the eyes (as in the case of head-mounted displays) or physically separate the two images displayed on a single screen, which is at a certain distance from the eyes. In the last category, the principle is based on a technical device that separates the images so as to show only the desired image to each eye. The classification given below is thus based on the place where the images are separated and the type of device used.

4.3.2 Separation at the screen level

Autostereoscopic flat screen with lights

The screen is a unit of two planes: One is an LCD screen and the other one, right behind the first, is made of thin illuminated columns separated from each other by dark zones. There is one light column for two columns of pixels of the LCD screen. Each light column is spaced out in such a way that an observer in a good position sees

---

[2]AR: Augmented Reality.

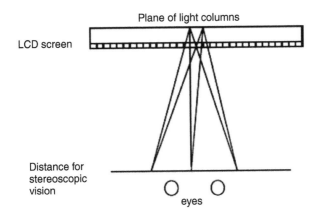

Figure 4.2 Principle of vision with auto-stereoscopic screen with lights.

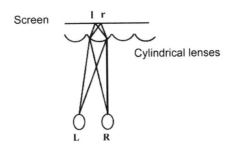

Figure 4.3 Principle of vision with auto-stereoscopic screen with lenticular network.

it through the even pixel column with his left eye and the odd pixel column with his right eye. In this device, the pixels can be seen only if they are lit up (Figure 4.2). By displaying the left image on even columns, the left eye of the observer sees the right image and similarly the odd columns display the right image to the right eye.

Autostereoscopic flat screen with lenticular network

The screen is covered with a lenticular network: A plane made of semi-cylindrical lenses with axes parallel to the columns of the screen. Their optical characteristics imply that each eye of the user sees different pixel columns, corresponding to two different images. For each eye, there are 1, 2 or 4 pixel columns behind each cylindrical lens. If there are more, the observer sees the continuous scene from different points of view by slightly moving his head. In this case, the vision is more similar to natural vision. But this change of point of view is possible only in a small portion of space (Figure 4.3). Various types of screens have been tested. Flat screens are currently used because it is easier to align the pixel columns with the cylindrical lenses. This is the main technical difficulty of auto-stereoscopic screens, mainly large size screens.

The main advantage of autostereoscopic screens is that the user does not need to wear any eyeglasses (very useful in public places for an immediate vision). However, as a result of the possibility of multiple points of view, reducing the resolution of images at least by 2 (or 4 or 8) is the main inconvenience. The moiré of the auto-stereoscopic screen degrades the image quality a little. The main disadvantage is that the observer must be in a specific zone, relatively restricted, so that both of his eyes perceive a good image.

Separation by colorimetric differentiation

The old and well-known procedure uses two-tone eyeglasses (anaglyph procedure). Each eye sees only one of the two images of different colours, red and cyan (complementary colour of red), thanks to the two filters on the pair of eyeglasses. Barco Company has recently updated this procedure (Infitec procedure). Each eye of the observer receives a different filtered image on the Red, Green and Blue (RGB) components of the colour spectrum. This can be achieved by putting on a pair of eyeglasses having two different colorimetric filters and by providing two video projectors, displaying the images on a large screen.

Separation by eyeglasses with electronic shutter (active eyeglasses)

These eyeglasses have two liquid crystal screens which shut each eye alternately 50 or 60 times per second, while the monitor displays images at a frequency of 100 or 120 Hertz. The normal frequency of the monitor is thus doubled to avoid flickering of images on each eye (25 images per second). A procedure is used to synchronise the shutters with the monitor's display: The synchronisation is transferred by infrared connections.

Separation by polarising eyeglasses (passive eyeglasses)

The technical principle consists of using light polarisation. A liquid crystal screen is placed in front of each of the two video projectors, which allows polarisation of light at a different time for each pair of stereoscopic images sent by the two video projectors. With eyeglasses having two filters that transmit only the desired image to each eye, the observer sees the images in relief. They are lighter than the active eyeglasses and do not require battery power supply. Polarisation of images is either a cross polarisation at 90° (which has the lowest performance and the lowest price for eyeglasses) or a "circular" polarisation. The normal frequency of the monitor is doubled to avoid flickering of images on each eye.

4.3.3   Large screen projection systems

To simulate a large field of vision to the user, some devices are based on video projectors displaying images on large passive screens. The images are often projected via mirrors to restrict the overall size of the system. But a new, less bulky technology is being developed, which will compete with video projection: the active flat stereoscopic screens, developed from flat screens (monoscopic) of a television. In both the cases, the stereoscopic technologies used are the ones using eyeglasses for image separation. All techniques of stereoscopy based on a separation of images by eyeglasses are possible. The number of video projectors used to display the images on the screen can vary from product to product. As the number of projectors per screen increases, the quality of

Figure 4.4 Reality Center by SGI, copyright photo SGI.

images improves, but at the same time, the complexity of the system also increases. If more than one projector is used for each screen, it is absolutely essential to provide for a system that ensures the continuity of contiguous images. There are two solutions possible: We can either juxtapose the projected images or align them perfectly or we can overlap them a little. In the second case, it is necessary to use an edge blending software or hardware to make these overlapping zones invisible to the eye. Several technologies are used: LCD (Liquid Crystal Display), DLP (Digital Light Processing), LCOS (Liquid Crystal On Silicon), etc.

The "immersive rooms" have a large flat screen or a semi-cylindrical screen. In case of a semi-cylindrical screen, the field of vision covered for the observer at the centre is in general 160° to 180°. These configurations are generally made up of a set of video projectors (at least three), making it possible to display a large, high-resolution image. Using several juxtaposed video projectors requires using the edge-blending technique to create a single image. The structures of immersive rooms are generally quite similar. The difference is mainly in their size and thus the number of seats. Another technical difference lies in the type of projection used: direct projection or rear projection? In the latter case, the observer can sit relatively closer to the screens, without casting his shadow on the screen (Figure 4.4).

CAVE are probably the most widely-known solutions using large screen projection. They are also the most expensive and the most difficult to set up and maintain. They are in the form of a cube-shaped room of about three meters. Stereoscopic images of the virtual world are projected on a number of surfaces ranging from four (three walls and the ground) to six for some configurations. This configuration creates a good sensation of immersion thanks to the layout of screens which gives the user a large field of vision, the stereoscopic visualisation and the display shown according to the head movements captured. Tracking the head helps calculating the images to be projected

Figure 4.5 CAVE with superior image quality (70 million *pixels*), copyright photo Renault.

on the screens as per the user's position, which in turn helps to continuously match his point of view with the camera in the virtual environment. By turning around an object, the user can change his point of view and examine it from any angle. If we want rear projection on the ground (to avoid creating the user's shadows), a transparent and sturdy screen would be required to bear the weight of a number of persons (a very expensive solution). The main differences between a CAVE and a VR headset: a CAVE have higher resolution and a higher field of vision than a VR headset and the user's body is always immersed in the virtual environment with a CAVE (Figure 4.5).

4.3.4 Wearable visual interfaces

Thanks to the successful in virtual reality, a new type of device has been developed for vision: VR headset (see the following chapter).

4.4 STEREOSCOPIC RESTITUTION OF VISION

It is about creating two images on a flat surface, each corresponding to each eye with the eyes seeing only one image due to a technical device. For each image, a point in space corresponds to a point on the flat surface. The distance between these two points is called parallax. The parallax creates binocular disparity in human visual system that gives a stereoscopic effect of depth with each eye receiving an image similar, but not

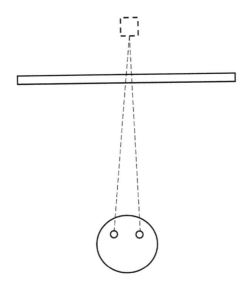

Figure 4.6 Positive parallax makes visualising objects behind the screen possible.

identical, to that of a real spatial vision. We emphasize that the images created are not identical and the images observed by the eyes are only plane images obtained by geometric projection of solids in space. These plane projections induce geometric and psychophysical constraints in plane images that spatial vision does not induce. We will explain these constraints in chapter 8. The value of the parallax varies according to the distance between the object and the screen. We have the following classification:

Positive parallax (not crossed)
When the object observed is located virtually behind the screen, the parallax is positive. The left and right homologous points are respectively displayed to the left and to the right (Figure 4.6).

Zero parallax
When part of the virtual object is located at screen level, the homologous points have zero parallax. They are displayed in the same places (Figure 4.7).

Negative parallax (crossed)
When the object observed is located virtually in front of the screen, the parallax is negative. The left and right homologous points are displayed to the right and to the left respectively (Figure 4.8).

As we have previously mentioned, the parallax should be small so as not to create great difficulties for stereoscopic vision. Otherwise, the brain either cannot merge the two images or it does it with a certain strain. The objective is therefore to create stereoscopic images with a low parallax while maintaining an effect of depth.

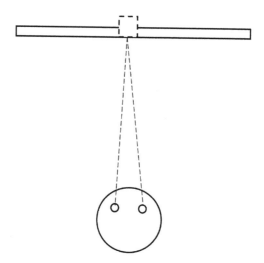

Figure 4.7 Zero parallax making visualisation of objects at screen level possible.

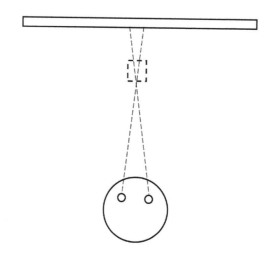

Figure 4.8 Negative parallax makes visualising objects in front of the screen possible.

The geometric reasoning is the same with a VR headset. The screen is close to the user's eyes but the focal length is around one and two meters (example for Oculus DK2: 1.3 meter): the distance between the user's eye and the virtual stereoscopic images, where the eyes accommodate thanks the optic lens (see chapter 5).

Chapter 3 on human senses explains in detail the human visual system and the perception of depth in monocular vision and in binocular vision. In a virtual reality application, every designer should ask the question: is stereoscopic vision indispensable? If the choice of the technical device allowing the display of stereoscopic images is important, it is not the only problem to be resolved for creating a 3D effect. Many people think that once the hardware is chosen, the creation of 3D images is a relatively simple question to be dealt with. We would like to show in the following paragraph, chapter 8 and chapter 9 that the configurations of algorithms for computer-generated images or parameters of cameras for real images are to be chosen wisely. For 3D images, compromises must be made. They are based on geometrical, psychophysical and cognitive considerations. Ignorance of these criteria in the past has led to the complete failure of projects using stereoscopic images. It should be remembered that 3D vision depends on the technical method implemented as well as on the visual system of the observer.

Chapter 5

VR headsets

5.1 INTRODUCTION

The objective of a VR headset is to provide a stereoscopic vision using two small screens, a large field of view corresponding to the visual field of the user and an immersion of the eyes. There is no ideal and universal visual interface but the main objective of a VR headset is to be a visual interface which must possess metrological characteristics that correspond to the maximum capacities of the human visual system. We, however, have a very long way to go to reach this goal. This visual interface should thus offer four additional capacities over an ordinary screen:

- *large horizontal and vertical fields of vision* corresponding to those of our eyes;
- *stereoscopic vision in the entire binocular field of vision*;
- *high graphic resolution* using all the performances of monoscopic and stereoscopic acuities;
- and a *gaze immersion* in the virtual world. Achieving this is possible only if the visual interface is a VR headset which is connected to a sensor locating the orientation of the operator's head (see section 5.5).

One of the major difficulties of VR headset is making it possible to observe the images on screens that are very close to the eyes, and for this we need to insert optics between the eyes and the screens. It is necessary to fully determine the geometry and the optics of the VR headset to be able to calculate and display the stereoscopic images correctly on the two screens.

These devices also have a location sensor tracking the user's head to display images corresponding to the orientation of his eyes. The direction of the eyes is by definition the projection of the axes of the eyes on the scene. It is thus the result of the orientation of the body, head and eyes. Hence, the images displayed in a VR headset correspond to the orientation of the head and not the direction of the eyes, as no VR headset marketed to date can detect the movement of the eyes, except some VR headsets with eye tracking (see chapter 7). In most of the cases, however, it is the direction of viewing. In fact, visual information is captured in the following manner: First, the eyes rotate in their orbit and then the head rotates slightly to put the eyes in the middle of their orbit. The VR headsets come with a 6 Degrees Of Freedom

(DOF) tracker or only with 3DOF (directions) in less expensive devices (see section 5.5). It is not always necessary to use six degrees of freedom, because this may depend on the requirements analysed at the level of sensorimotor I^2. If we move the head, a fixed object is seen from different angles and the dimensions of its images on the two retinas vary. However, our brain effortlessly perceives this object as fixed. The VR headset must not disturb this visual perception of stationary objects. The images displayed on the screens of the VR headset should thus give the same visual sensations as the ones we get when we see the real world, or at least be similar to the same. This not only depends on the characteristics of the VR headset, but also on the quality of the head location sensor and the algorithms calculating the images to be displayed. Using a VR headset thus requires accurate positioning of the two screens in space as well as of these screens with respect to the eyes. Several problems are to be identified:

- Positioning the VR headset on the head: the display should be positioned correctly on the user's head, i.e. we must be sure that we are displaying the images of the correct point of view;
- Location of the user's head: finally, the position and the orientation of the VR headset in space must be known precisely so as to generate good images.

Theoretically the field of view is 360 degrees, considering the movements of the head. When the head is at a fixed position, the field of view of a VR headset is not as large as it would naturally be (roughly, it is rarely more than 100 degrees compared to 180 degrees in a horizontal plane), it nonetheless gives an impression of visual immersion. In natural vision, the perception is clear only in the central field. If we observe an object on the border of the visual field, we instinctively turn our head to see it better. The VR headsets are thus less effective in drawing the user's attention to the side, as their field of vision is very narrow. In certain VR headsets with very restricted field of view, the user feels that he is looking through a tube, which certainly is an unpleasant feeling.

Besides, a stereophonic sound or better a spatial sound (3D sound) is also offered as these devices have two headphones. 3D sound: a group of sound effects that manipulate the sound produced by headphones or stereo speakers. This involves the virtual placement of sound sources anywhere in three-dimensional space.

One or two small screens are used to design a VR headset. The main difficulty is to manufacture screens with a high density of pixels. If we want to increase the field while maintaining a sufficient screen resolution, we will have to display a huge number of pixels which is not compatible with current technology. The average value (1 minute of angle) of visual acuity must be referred to in order to judge the resolution quality of a screen or VR headset used in a VR device. In the 1990s, the development of VR headsets was progressing at a very slow pace. The technical difficulties, especially in terms of optics and of miniature screens, was enormous and the market for this type of visual interfaces was very limited. The demand was thus not driving the development of VR headsets. But thanks to the large smartphone market, miniature screens with high-resolution are available now. The following paragraphs present the different solutions. The VR headsets can be divided into a number of groups based on their structure.

Figure 5.1   Google Cardboard (courtesy of Google, Inc.).

5.2   DIFFERENT TYPES OF VR HEADSET

5.2.1   VR headsets designed for a smartphone

It is interesting to design a VR headset for a smartphone because many people have this device and all smartphones have a tracker and a small high-resolution screen. This type of VR headset is low-cost because the consumer has to buy only a frame made from cardboard. We call these VR headsets: "smartphone-based headsets". Obviously, their performance is highly dependent on the smartphones used: the screen resolution, the latency, the screen contrast, etc. In the category of smartphone-based headset, there are two types of device:

– The devices that cannot be fixed on the user's head and which must be maintained with the hands;
– The devices that can be fixed on the user's head.

5.2.1.1   Cardboard VR headsets

This type of device uses a pair of plastic magnifying lenses and a sheet of cardboard, using a standard smartphone as a screen. This kind of material has no adjustment to suit different users' morphology. They offer limited interactivity, most suited for watching 360-degree video (Observational VBP). Most people refer to the best known "Google cardboard" (Figure 5.1) but the idea has been known before Google branded it (see chapter 7).

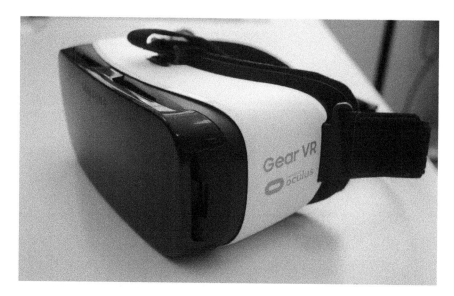

Figure 5.2 Gear VR.

5.2.1.2 *Smartphone-based headsets with head band*

These devices are also smartphone-based but are more robust than the models presented above. They are relatively light with an adjustable headband of a few centimeters wide. They offer the user the ability to adjust one or more parameters such as intraocular distance. The best known is the Gear VR (Figure 5.2). We know that it is very difficult to design and manufacture a VR headset and thus the smartphone-based headsets are secondary solutions. But the main advantage of using a smartphone-based one is that almost all people have one. Another advantage is the lack of cable connecting the computer to the VR headset but the performances are limited (see chapter 7).

5.2.2 Headsets which are intrinsically designed for VR

The best-quality VR experiences can't be powered by a mobile phone. The headsets which are designed intrinsically for the VR applications have better features. They are also generally more comfortable and better at blocking outside light. These devices are different than previously presented devices by the nature of their hardware in order to provide better performances. These VR headsets use cables which restrict the movements of the user. One of the big features of these headsets is the ability to move or even walk through space. Most of the current devices have a limited field of view (around 100 degrees).

But the Star VR (Figure 5.3) offers the widest field of view of all the VR headsets available on the market with 210° horizontally and 130° vertically. It is the first "full-face VR headset". This headset was designed in 2013 under the name InfinitEye by a small French team. The company went on to be acquired by Starbreeze Studios. The view inside of Star VR is truly impressive. It feels like putting your head into the

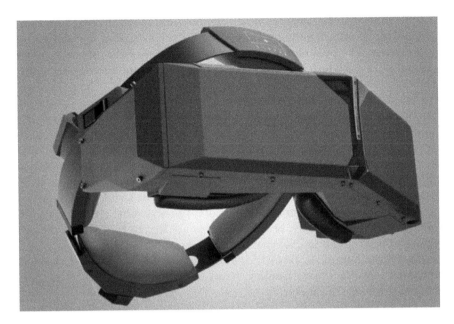

Figure 5.3 Star VR (courtesy of StarBreeze).

virtual world in a much more significant way than other headsets. Furthermore, some VR headsets incorporate eye tracking (see chapter 7).

5.2.3   AR headsets

The AR headsets are devices dedicated to Augmented Reality (AR). There are two categories for this type of device: Some AR headsets are based on conventional screens "video see-through", and some devices are based on semi-transparent screens "optical see-through. A "Video-see through" device displays video from camera which films the real scene and which is outside the headsets. "Optical see-through" systems combine computer-generated images with the view of the real world through a semi-transparent mirror. Some AR headsets have complete occultation of the field of view and other AR headsets have partial occultation (see chapter 7). An advantage of using an AR headset: the user sees partially the real world, meaning he is even more stabilised thanks to his peripheral vision which lies in the real environment, as technically it is not possible to display peripheral-vision images in AR headsets. We will talk about this subject in chapter 9.

5.3   THE DESIGN OF OPTICAL SYSTEM

The major difficulty of the VR headset is to design the optical device. A solution is to insert two lenses between the eyes and the screens. The technical constraint is to add an optical device with a large field of view for the accommodation of eyes on screens that are very close to them. It is the main constraint restricting the horizontal visual field to a maximum of 100–120 degrees if the solution is only to insert two lenses between the eyes and the screens. Another solution is to use a Fresnel lens in order to increase the field of view. The field of view of a "full-face VR headset" (Star VR) requires a complex set of lenses which, in this case, combine normal and Fresnel elements.

There are several sources of errors in the optical models. The optical systems involve defects of optics. These create aberrations (spherical aberration, astigmatism, distortion, etc.). These defects must be taken into account in an optical model to be able to correct the images using a software package. The VR headset manufacturers supply this software package. The interpupillary distance varies from one person to another. To avoid these errors, it is necessary to provide for either a mechanical setting or optics with sufficiently large outlets so that anybody can use it. The lenses of the VR headset are responsible for mapping the up-close display to a wide field of view, while also providing a more comfortable distant point of focus. One challenge with this is providing consistency of focus: because the eyes are free to turn within the headset, it's important to avoid having to refocus to prevent eye strain.

Various techniques have existed for AR headsets. Most of these techniques can be summarized into two main families: "Curved Mirror" based and "Waveguide" or "Light-guide" based (see chapter 7).

5.4   DISPLAY SCREENS

5.4.1   Current display screens

There are several characteristics in order to judge the quality of the display screens. The more important is the resolution[1] which is defined by the number of pixels per inch. The pixel density is also a very important criterion. As we can see, the visual acuity is about 1 minute of angle. The average value of visual acuity must be referred to in order to judge the resolution quality of a screen. To follow this characteristic, a 25 cm wide screen positioned a distance of 35 cm of the eyes should display 2500 pixels horizontal. These numeric values are established for a screen giving a narrow horizontally field of view (40 degrees). But the horizontal fields of view of most VR headsets is about 100 degrees. With this characteristic, the screens of VR headsets should display a huge number of pixels (about 6000 pixels horizontally) which is not compatible with current technology. The resolution is weak in all current VR headsets (see chapter 7). For a "full-face VR headset" (horizontal field of view: 210° and vertical field of view: 40°), it should have the definition: horizontally $210 \times 60 = 12600$ pixels and vertically $140 \times 60 = 8400$ pixels, total: about 100 million pixels! In order to compare the resolution of the VR headsets, it should better indicate the number of horizontal pixels per degree (ppd). This number is not dependent value of field of view.

---

[1]The resolution measures the smoothness of the details on a screen. It is defined by the number of pixels per inch. The definition is the number of all pixels that are displayed on the screen. It is defined by the number of horizontal and vertical pixels, for example: full HD = (1920 × 1080).

The quality of a screen depends on resolution but also luminance and contrast of the screen. Luminance is the amount of light energy emitted or reflected from a screen in a specific direction. Luminance is measured in candelas per square meter. The luminance that we can observe in nature covers a range of 10^{-6} to 10^6 cd/m^2 (a moonless night has a luminance of approximately $3 \cdot 10^{-5}$ cd/m^2 and the snow in bright sunlight gives a luminance of approximately $16 \cdot 10^3$ cd/m^2) and the human visual system can accommodate a range of about 10^4 cd/m^2 in a single glance.

The contrast (exactly: luminance contrast) is the ratio between the higher luminance and the lower luminance that defines the feature to be detected. This contrast ratio is used for high luminance and for specification of the contrast of screen. For example, if it is possible to display on a screen a weigh colour at 200 cd/m^2 and a black colour at 0.25 cd/m^2, the contrast ration is: 800:1. The luminance of black colour cannot be zero because a screen is never perfect. A correct screen displays a black colour under 0.1 cd/m^2 and a wrong screen displays a black colour above 0.3 cd/m^2. A screen has a better visual quality if its contrast ratio is higher compared to another screen, the two screens being equal resolution.

The screens of VR headsets come from screens of smartphones. The actual definitions of best screens are Full HD (1940 × 1080 pixels) and QHD (2560 × 1440 pixels). There are several types of display technologies: LCDs (which act as valves to allow varying amounts of the backlight through to the viewer) or AMOLED/OLEDs (a different approach that uses "active-matrix organic light-emitting diodes" to directly emit light). LCD screens start with an always-on backlight; this technology requires light to create black, white, and colours. The liquid crystals then make it possible to modulate the quantity of light transferred. OLED does not require any light to produce black, only white and colours. Therefore, it's considered battery-saving (since it requires no energy to create black) and can produce inky blacks. What makes the two differing technologies more confusing is that there are multiple versions of each. For instance, IPS (in-plane switching) is a type of premium LCD technology that's touted for its wide viewing angle and clearer picture. An AMOLED display involves an active matrix of OLED pixels generating light upon electrical activation that have been deposited or integrated onto a thin-film transistor (TFT) array in order to control the current flowing to each pixel. Each screen pixel is actually composed of red, green, and blue subpixels that can turn on and off in combination to create any supported color combination. But there is also the PenTile screen that in fact, uses fewer red and blue subpixels than it does green. The basic PenTile structure is the RGBG matrix. In RGBG PenTile displays there are only two subpixels per pixel, with twice as many green pixels as red and blue ones. There are also other subpixel arrangements.

5.4.2 Future display screens

New technologies are developed. The first such technology is the "Light Field" display. Light field displays are displays that use an array of tiny lenses to display light focused at many depths simultaneously. This light-field-based approach to near-eye display allows for dramatically thinner and lighter VR headsets capable of depicting accurate accommodation, vergence and binocular-disparity depth cues. The near-eye light-field displays depict sharp images from out-of-focus display elements by synthesizing light fields that correspond to virtual scenes located within the viewer's natural accommodation range. Near-eye light-field displays support continuous accommodation of the eye throughout a finite depth of field. As a result, binocular configurations provide a means to address the accommodation-convergence conflict that occurs with existing stereoscopic displays.

In Augmented Reality, we have to mix light from real and virtual objects. The light from the real world will naturally be focused at a variety of depths. However, the virtual content is focused at a fixed, artificial distance dictated by the optics with a standard screen. Virtual objects do not look like they are really part of the scene. They are out of focus when we look at real things at the same depth. It is not possible to move the eye fluidly across the scene while keeping it in focus, as we do normally. It is the conflicting depth cues between accommodation and vergence. Light field displays allow the user to focus naturally on the display, and solve the problem described above for Augmented Reality. Light field displays have a resolution decrease in order to give adequate depth precision. It is the challenge to manufactured light field displays with high resolution. The light Field displays will be used in the Hololens AR headset of Microsoft. Like Microsoft's HoloLens, Magic Leap is using waveguides to superimpose 3D images over real world objects. In Hololens, the waveguides work by creating a light field, but via projection zigzagging across the RGB waveguides. Magic Leap doesn't like describing its technology as lenses, instead opting to call it a "photonic light field chip".

A Virtual Retinal Display (VRD) utilizes photon generation and manipulation to create a virtual image that is projected directly onto the retina of the eye without creating a real or an aerial image that is viewed via a real screen, mirror or optics. The user sees what appears to be a display floating in space in front of them. The VRD includes a source of photons, which are modulated with image information and scanned in a raster type of pattern directly onto the retina of the user's eye. The photon generator may utilize coherent or non-coherent light. The virtual retinal display may also include a depth accommodation cue to vary the focus of scanned photons rapidly so as to control the depth perceived by a user for each individual picture element of the virtual image (As a result, no accommodation-convergence conflict). Further, an eye tracking system may be utilized to sense the position of the user's eye. But it is a difficult challenge to track the eyes with very low latency (see the book "Displays, Fundamentals & Applications", by Rolf R. Hainich and Oliver Bimber[2]).

5.5 HEAD TRACKING SENSOR

5.5.1 Introduction

This paragraph describes the sensors that capture the real-time location of any object to know its position and (or) to follow its path. In virtual reality techniques, the object to be located is often a body part, the entire body or an object handled by the user. When they are used in virtual reality, we call them "trackers". These sensors are designed to measure only the spatial position of a moving reference point which is linked to the head or the hands observed, with respect to a fixed reference point.

---

[2]ISBN 978-1-56881-439-1, publisher CRC Press.

The position of a reference point linked to an object can be determined perfectly if the tracker measures the 6 parameters (3 angles and 3 distances) connected to the six degrees of freedom (or 6 DOF) of the reference point. These 6 parameters are defined by the matrix of geometric transformation between the reference point in the environment and the reference point linked to the object. This transformation includes two motions – rotation and translation. The parameters can be defined in different ways and generally the measuring systems let the user choose from various representations. Regarding the rotations, it is interesting to utilise the quaternions and not the Euler angle in order to avoid the singularities. Nevertheless, designing and creating a reasonably-priced location sensor that provides six extremely accurate measurements in real time (at a frequency higher than 100 Hz), with no operating constraints, is a very difficult technical challenge. We will thus concentrate on only the sensors that locate an object in a restricted area, such as the area around a computer or a part (a few centimetres to a few decametres). Various principles of physics are currently in use. Each of them has at least one advantage over the others. Some of the physical phenomena used are mechanic or optic. A major technical challenge for VR headsets is to make head tracking as good as possible. The metric that matters is called "motion-to-photon latency". This is the time that it takes for a user's head motion to be fully reflected in the rendered content.

5.5.2 Mechanical trackers determining an orientation, speed or acceleration

Many types of mechanical trackers are covered in this section as they are based on mechanical principles that provide indirectly, the measurement of *only one degree of freedom*. It is therefore necessary to combine them with each other tracker to create 3 or 6 DOF location sensors. Some of these trackers were developed for aircraft and missile guiding systems. Miniature trackers with an affordable production cost were designed using similar principles for the video game market. These sensors are used for smartphones and VR headsets. The different types of mechanical sensors used in virtual reality are inclinometers, gyroscopes and rate gyros and accelerometers.

Inclinometers: The earth is the source of gravitational field for this type of tracker. When the inclinometer is not moving, it indicates its angular position with the vertical direction of the place. But when it is moving it indicates the direction of the total external force on it, caused by the acceleration of the movement and the earth's gravity.

Gyroscopes and rate gyros: The most common gyroscope is the one based on the mechanical principle of a rotor, spinning at a high speed, whose axis maintains a constant direction. However, other physical phenomena are used to measure angular rotational velocity.

Accelerometers: Accelerometers, as their name suggests, are based on the measurement of a force coming from the acceleration of a mass. This measurement is detected by the piezoelectric, piezoresistive, strain gauge or capacity variation principle. The measurement of location of a moving object is obtained by double integration, which causes a quadratic increase in position errors. Acceleration due to gravity must be taken into account, for example in an inclinometer. With VR headsets designed from a smartphone, tracking the user's head orientation is done using onboard sensors of the smartphone. Only the rotation of the head is tracked using the sensors embedded in the smartphone!

5.5.3 Optical trackers

There is no standard tracker in this category. The trackers are generally grouped on one of the following two principles: outside-in or inside-out. In case of outside-in, the sensors are fixed and the markers are embedded on the VR headset. The difficulty of this principle is that the reference points move relatively less than the rotation of the VR headset, which results in less accurate measurements in rotation but more accurate measurements in translation. In this condition, we must equip the room with several cameras. In case of inside-out, the markers are fixed and the sensor is embedded. All the smartphone-based headsets are a camera but this sensor is not often used in order to track the headsets. The difficulty of the inside-out principle is less accurate measurements in translation. Currently, the two principles are used with the actual VR headsets (see chapter 7).

The trackers can also be classified on the basis of the technology used:

- Tracker working in the visible spectrum or in the near infrared spectrum;
- Trackers using passive markers or active markers.

Passive markers are generally small surfaces (disc, sphere, etc.) covered with retro-reflective paper. The advantage of passive markers is that they do not require any wires. On the other hand, they need a light source. In case of spheres, the light source is close to the camera's optical axis so that the reflected rays are captured by camera. Active markers are generally made of LEDs. Their main drawback is that they need power. Their major advantage is that they can be triggered one after the other, which makes it easy to identify them in the image. They are preferred in trackers operating on the inside-out principle.

For example, the ARTtrack system is marketed by the ART (Advanced Real-time Tracking) and used with CAVE (Figure 5.4). It follows the Inside-out Principle. A tracker system consists of at least two cameras (this number can go up to eight), a computer and markers. The number of cameras depends essentially on the volume of work. The markers are in the form of disks or retro-reflective spheres. The tracker reflects the location for the isolated markers and gives the location and orientation for the mechanically rigid structures (rigid bodies). These structures are nothing but a group of cleverly arranged markers. In a typical application, the rigid bodies are mounted either on a VR headset or a pair of stereoscopic glasses, to track the point of view. The proclaimed performances in terms of accuracy are 0.4 mm in translation and 0.12 degrees in rotation, for a maximum frequency of 60 Hertz.

The main disadvantage of these optic trackers is that the line of sight from the camera to the marker cannot be blocked. Nevertheless, this point has been taken into consideration in the latest systems, mainly due to the redundancy of cameras and markers. The HTV Vive uses several sensors including gyroscope, accelerometer and laser position sensors. The motion tracking comes with two wireless infrared "Lighthouse" cameras, which are placed in the corners of a room, and follow the headset's 37 sensors (70 in total, including both motion controllers). The emitted laser

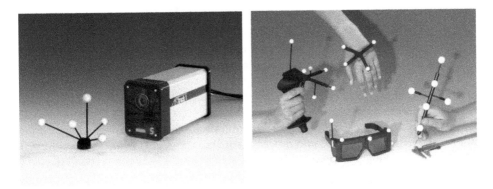

Figure 5.4 The ART track system (courtesy of ART GmbH).

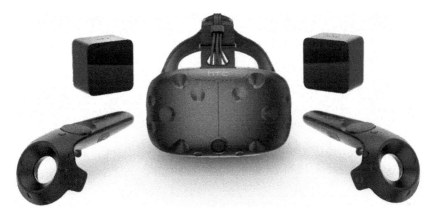

Figure 5.5 The optical tracker with two controllers and the HTC VIVE (courtesy of HTC & Valve).

sweep structured light lasers and the beams reflect on several diodes positioned on the HMD. This reflection is captured by the cameras and analyzed to compute the orientation and the position of the HMD within a diagonal area of up to 5 m. The "Lighthouse" base stations track the user's movement with sub-millimeter precision.

5.6 THE ERGONOMIC DESIGN

An ergonomic problem is well known: the weight of the VR headset can induce an increase in physical symptoms when the subject wears a heavy helmet for too long. However, given that the equipment is becoming increasingly sophisticated, this problem can potentially be avoided by wearing a lighter helmet. VR headsets and their fastenings must be adjustable to any head shape using fairly effective technical solutions. It is quite amusing to see on demonstration videos, a person removing his hands from the VR headset to manipulate a virtual object and then replacing nimbly his hands on the VR headset, which looks like it needs help to stay on his head. Another constraint is the heat users may feel on their face. Poor ventilation in the VR headset can also cause fogging on the lenses on some models.

The cables connecting the computer to the VR headset can also hinder certain of the person's movements. Cables are set to remain for some time on high-performance VR headsets, while those used with smartphones do not have any. One solution for a person who is standing and is required to swivel around could be to wear a computer on a belt – a solution particularly favoured by VR headsets used for Augmented Reality. Designers need to take into account the cables and the potential hindrance to the person's movements. Cables could descend from the room's ceiling before reaching the VR headset to be less restrictive.

The problem of the discomfort caused by the weight needs to be tackled. Using a VR headset is intrusive, in the sense that it is heavy, its movements can be more or less restricted due to the video cables, and the head's centre of gravity is moved. Several questions thus need to be answered:

- Does the weight of the VR headset have an effect on the head's position? The shift in the centre of gravity of the head and display can make the user look for a new, less tiring position;
- Are the user's movements the same with or without the VR headset?

It is thus necessary to try to estimate the bias caused simply by wearing a VR headset, even before thinking about the vision, if the posture of the user's head is important for the application (for example, study of the visibility of a road scene from inside a vehicle). So as not to disturb the movements of the head, the VR headsets need to be symmetrical, as light-weight as possible and should have a centre of gravity towards the front of the head so as to take support of the dorsal muscles of the neck that are stronger and less prone to fatigue.

There are several arrangements for individuals using VR applications, the main ones are as follows:

- the person is sitting on a fixed seat;
- the person is sitting on a seat swiveling 360°;
- the person is sitting on a motion simulation platform;
- the person is standing and is motionless;
- the person is standing and moving around the room;
- the person is walking on a treadmill;
- the person is stretched out on a swing seat.

The ergonomic constraints will differ according to the arrangement selected. Some solutions are explained in chapter 9.

If we want to study the quality of a VR headset as a visual interface, asking the user whether he can see the images is not enough. The characteristics of visual interface depend on VR headset but also on the computer imaging software and on your visual system! We have offered a number of tests related to the professional use of a VR headset. It is necessary to validate the sensorimotor immersion and interaction (I^2) before studying cognitive and functional I^2.

For location of the head-mounted display, the two major problems are:

- Is the accuracy of location of the head sufficient for the application?
- Is the latency time taken by the sensor and the creation of images acceptable?

For vision, the two major problems are:

- The compatibility of the visual fields in the VR headset in the virtual environment with the visual fields used in reality (horizontal, vertical and overlapping zone);
- Usefulness, validity and constraints of stereoscopic vision for the application.

Chapter 6

Interfaces used with VR headsets

6.1 INTRODUCTION

In this chapter, we do not present all VR interfaces but only the main devices which are exploited with VR headsets. Some behavioural interfaces (user interface) are designed for any type of visual restitution while others are specifically designed to be used with VR headsets. Several handled controllers are designed and provided by the manufacturers of VR headsets. In this chapter, we present the following interfaces:

- Tracked handheld controllers;
- Locomotion interfaces: VR treadmill;
- Motion simulator.

If we observe an object on the border of the visual field in natural vision, we instinctively turn our head to see it better (see chapter 5). The VR headsets are thus less effective in drawing the user's attention to the side, as their field of vision is very narrow. But we can exploit 3D sound technology with two headphones to draw the user's attention to the side. In this chapter we do not explain the software to create 3D sound which is correctly developed now.

6.2 TRACKED HANDHELD CONTROLLERS

There are several types of devices which are used to interact with a virtual environment. There are some common behavioural interfaces such as joystick, gamepad and so on, but others are more complex, for example the force feedback interfaces which apply forces to the user's hand who handles a virtual object. There are also the tactile interfaces in order to stimulate the sense of touch: a tactile device can reproduce as truly as possible the tactile parameters, such as the texture, roughness, temperature, and shape[1]. Force feedback interfaces are very expensive and are mainly used for some professional VR applications. Tactile interfaces are most often low-cost if these devices

---

[1] Tactile interfaces: a state-of-the-art survey, M. Benali-Khoudja, M. Hafez, J.M. Alexandre and A. Kheddar, ISR International Symposium on Robotics, Paris, March 2004.

Figure 6.1 The wireless controllers of HTC Vive (courtesy of HTC & Valve).

just provide simple tactile stimuli, for example a click. It is more difficult to produce a wide variety of tactile feelings with vibrotactile patterns.

VR headset technology is fairly modern, it is constantly evolving and has not been standardized yet and in contrast, game console controllers and joysticks have been standardized for more than 25 years. The Wiimote brought a new way of controlling by adding orientation tracking and the Kinect enabled positional and body tracking. The VR headset manufacturers (Oculus, HTC Vive, etc.) currently invent new controllers and head-trackers. The controllers of new VR headsets are essential for the user's interaction because they are specially designed for VR application with headset. Normally, a controller system consists of a pair of handheld devices for the left and right hand, each containing a joystick, buttons and triggers for grabbing or shooting (Figure 6.1). The controllers are tracked in 3D space by the system which uses the same principles of head tracking.

The controllers enable to manipulate objects in the virtual environment with high precision and low latency. The combination of tracking and an integrated grip button makes it easy to pick up and manipulate objects. The controllers are wireless so one can move and interact freely in the virtual world. Some controllers include tactile feedback on buttons or triggers. The controllers enable to manipulate objects with colocalisation. Colocalisation reflects the fact that the visual space matches the manipulation space (Figure 6.2). The advantage of colocalisation is that it increases the manipulation efficiency and increases user precision. But it is possible to manipulate objects with a non co-localised hand via teleoperation (see chapter 9).

A dataglove is a device which tracks the fingers movements and knows the positions of the user's fingers. But without glove, it is possible to track the fingers of both hands. Fingers tracking is an optical technique that is employed to know the consecutive position of the fingers of the user. The Leap Motion controller is a small peripheral device which is designed to be placed on a physical desktop and can also be mounted onto a VR headset. Using two monochromatic IR cameras and three IR LEDs, the device observes a roughly hemispherical area, to a distance of about 1 meter. The overall average accuracy of the controller was shown to be 0.7 millimeters. The smaller observation area and higher resolution of Leap Motion differentiates the device from the Kinect, which is more suitable for body tracking.

In order to touch and manipulate a virtual object, it is possible to use a haptic dataglove. For example, the "Gloveone" uses ten vibrotactile actuators and enables users to touch any virtual object that they can see on their VR headset or screen. But this device needs a finger-tracker device (the Leap Motion for example – Figure 6.3).

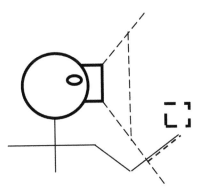

Figure 6.2 Manipulation objects with colocalisation (Real environment: continuous line and virtual environment: dotted line).

Figure 6.3 The Leap Motion.

With this type of dataglove you can feel shape but no weight, because there are no force feedbacks. This type of device enables to interact with virtual buttons and elements, stimulate differentiate textures, receive haptic warnings and alert events, etc.

6.3   1D TREADMILL AND OMNIDIRECTIONAL (2D) TREADMILL

A first way of virtual walking is to use a standard 1D treadmill in order to walk while staying in the same place. The speed of walking is measured and controlled in order to compensate the walking movement. However, the problem with this solution is that the direction is imposed. The treadmill must be supplemented by a controller (joystick, buttons …) which enables to change the direction of virtual walking. But in this case, the trajectories are not the same as moving naturally because the user cannot turn

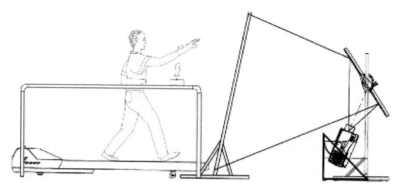

Figure 6.4 The 1D treadmill for the training simulator of SNCF society.

around instantaneously, for example: The 1D treadmill which is exploited with a large display in order to train the operating staff of French railway SNCF society (Figure 6.4). The users are not always comfortable when they walk on a treadmill, mostly wearing a VR headset. With a treadmill, the proprioceptive organs of the muscles, tendons and joints are able to be correctly simulated to reduce the sensorimotor discrepancies, while this is not the case for vestibular systems. The visual-proprioceptive discrepancy will be reduced in this case (overall proprioception) but the visual-vestibular discrepancy will always be present (see chapter 9).

An omnidirectional treadmill (or 2D treadmill) is another solution of virtual walking. It is a mechanical device that allows a user to perform locomotive motion in any direction. An omnidirectional treadmill gives free movement to its user in line with any virtual environment that makes use of the user's physical motion. There are three main types of technological solutions. A first solution is a 2D treadmill which features a flat, circular platform that has a slippery surface, which helps to facilitate smooth movements. The user walks and his feet slip on the platform while staying in the same place. The user is able to spin around while slipping on the platform. The platform itself has a ring support attached to vertical beams that acts as the safety harness, preventing falls and slips while being engaged in the virtual environment. This walking is not very natural but efficient and the 2D treadmill is quite low cost. The Cyberith Virtualizer[2] has a low-friction surface that enables you to walk and run. Its ring allows for vertical movements such as jumping and crouching, as well as a 360° axial rotation. The user just has to slip into custom overshoes. There are other "2D slipping treadmills": Virtuix Omni[3] with a concave platform which enables a quite smooth, natural gait and an immersive walking and running motion. Kat Walk[4] is an omnidirectional treadmill which has an input device in order to capture the user's body-motion and movement, analyze the data and maps it into virtual environment (Figure 6.5). The biggest difference between this treadmill and other existing 2D treadmills is that it

---

[2] http://cyberith.com/product/
[3] http://www.virtuix.com/
[4] http://www.katvr.com/

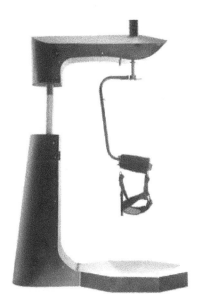

Figure 6.5 The Kat Walk omnidirectional treadmill.

doesn't have a ring or column surrounding the player, which was there to protect the player from falling. It uses a nest-like pouch, which will protect the user while allowing to move freely and safely without being constrained.

Unlike the "2D slipping treadmills", which are stationary platforms, other 2D treadmills are "active omnidirectional treadmills". In the late 1990s Professor Hiroo Iwata, University of Tsukuba, developed a walking system: Torus Treadmill is a locomotion interface equipped with special arranged treadmill. It can provide Infinite plane for creation of sense of walking. Torus treadmill consists of ten belt conveyers. Ten belt conveyers are connected side by side and driven to perpendicular direction. Torus Treadmill can provide infinite plane for walking (Figure 6.6). Some omnidirectional treadmills are marketed. For example, the Infinadeck[5] treadmill reacts to the user's movement, including their speed and direction, to keep them safely in the center of the deck. It works by driving a square set (about 1.5 m × 1.5 m) of parallel "mini-treadmills" to precisely simulate walking and running movements aligned with any virtual environment. This treadmill uses a looping belt powered by two motors which could propel a user standing atop it in the opposite direction he intends to walk. The treadmill includes a harness to detect directional input. The Omnideck 6[6] is an omni-directional treadmill. It features an optical tracking system that accurately monitors the user's movement in virtual space to maintain a sense of realism. The Omnideck 6 is driven by a responsive motor that allows to perform actions such as walking and running at speeds up to 2.2 m/s.

---

[5] www.infinadeck.com
[6] http://www.omnifinity.se/

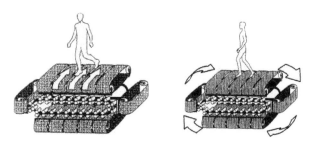

Figure 6.6 Movements on the torus treadmill (Professor Iwata).

Figure 6.7 The author running in a spherical omnidirectional treadmill.

The third type of technological solutions is designed with a sphere. For example, the Virtusphere[7] consists of a 10-foot hollow sphere, which is placed on a special platform that allows the sphere to rotate freely in any direction according to the user's steps. The platform tracks the body movements with motion sensors. It is easy to walk inside the sphere but more difficult to run because of the mechanical inertia of the sphere (Figure 6.7).

---

[7]http://www.virtusphere.com/

6.4 MOTION SIMULATOR

6.4.1 Introduction

The purpose of a motion simulator is to change the orientations and translations of user's body. But the motion simulator applies indirectly sensory stimuli of muscle proprioception and touch. A motion simulator is a device that creates the effect of being in a moving virtual environment, for example moving car. Motion simulators can be classified according to whether the occupant is controlling the simulator or whether the occupant is a passive user. The movement is synchronous with visual display: large screen or VR headset. This type of device is exploited with the dynamic driving simulator (see chapter 11). Motion platforms can provide movement on up to 6 degrees of freedom (dof): 3 translational dof and 3 rotational dof (roll, pitch and yaw). But other types of motion simulator have 2, 3 or 4 dof, for example a motion seat with 2 or 3 dof. The "full motion simulators" move the entire occupant compartment and can convey changes in orientation and the effect of false gravitational forces. The mind's failure to accept the experience can result in motion sickness (see chapter 8).

The design of motion simulator is a hard mechanical problem for the manufacturer. The weight of human is important and it is not easy to manufacture a low-cost motion simulator. These devices will always be expensive and so rarely will be used at home. The price of a motion seat simulator is about ten thousand dollars. There are different types of motion simulator:

–   Motion seats for single-user;
–   Motion flight simulators for single-user;
–   Full motion simulators for single-user or multi-user.

6.4.2 Motion seats for single-user

A motion seat (or motion seat simulator) enables to change only the orientations of user's body. The user is sitting on a seat or also standing on a surf simulator. These devices have controllers (joystick, wheel …) in order to control the moving in the virtual environment. The rotational accelerations on the body are often limited. There are sometimes low translation accelerations with the vibrations of the seat. The seat simulators often exploit electrical actuators. The display is a large screen in front of the user or a VR headset. This type of motion simulator is exploited mainly in amusement park, for example driving a car. The motion seats for single-user have more or fewer degrees of freedom: 2 dof for a motorbike simulator, 3 dof for a surf simulator, between 2 and 4 dof for a driving simulator, etc.

The simXperience[8] racing simulator includes tactile feedback technology in addition to 3 dof motion. It provides 250 motion updates per second to ensure that the user feels every detail of the vehicle suspension. It exploits a multi-dimensional audio based feedback software that utilizes one to eight bass transducers to provide the correct physics based vibrations at each corner of the simulator and throughout the simulator.

MMOne chair[9] is a new simulator that moves the user's body strapped to a chair at the end of an industrial robot. The user is wearing a VR headset. The control system

---

[8] http://simxperience.com/
[9] http://www.mm-company.com/

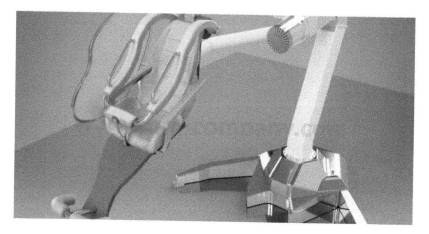

Figure 6.8 The MMOne chair simulator.

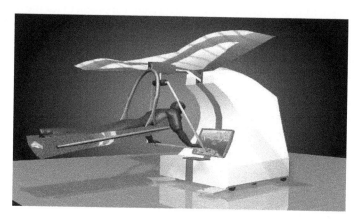

Figure 6.9 A motion flight simulator based on a hang-glider.

includes a joystick or an interactive wheel. The MMOne chair works on three axes and is able to rotate 45 degrees left and right, raise 45 degrees and revolve 360 degrees. The system offers a one millisecond response time with 20 millisecond position correction. Physical space required for the chair to operate safely is slightly more than four cubic meters (Figure 6.8).

6.4.3 Motion flight simulators for single-user

Another type of motion simulator is the motion flight simulator for single-user. Several flight simulators were designed for twenty years. Some motion flight simulators are based on a hang-glider (Figure 6.9) or paragliding. Other simulators are inspired by birds, for example the Birdly VR simulator[10]. The user wears a VR headset

---

[10]http://www.somniacs.co/

Figure 6.10 A dynamic driving simulator with 6 dof (courtesy of Renault).

to experience the immersive flying trip. The software simulates the bird's flying field of view while the user lies flat on a platform. The user commands his flight with arms and hands: their movements are directly correlated to the flapping of both wings.

6.4.4 Full motion simulators

The full motion simulators are mainly exploited for entertainment in amusement parks and for professional applications. For example, with a dynamic driving simulator (Figure 6.10), the ability to link a computer-based dynamic model of a system to

Figure 6.11 The Renault Ultimate simulator (courtesy of Renault).

physical motion gives the user the ability to feel how the vehicle would respond to control inputs without the need to construct expensive prototypes. They are also exploited for training of airline pilots. These simulators can provide movement on up to 6 dof and are commonly used in the field of engineering for analysis and assessment of vehicle performance. The visual display is often a large screen. These simulators exploit electrical motors or hydraulic motors, which are more powerful but more expensive. For example, the specifications of the standard Inmotion Simulation platform[11] are rotational acceleration: $150°/s^2$ and displacements: $\pm30°$ roll, $\pm30°$ pitch, $\pm45°$ yaw, $\pm32''$ heave, $\pm34''$ surge and $\pm 34''$ sway.

There are some complex full motion simulators with 8 dof motion in order to have the best driving simulations. For example, the Renault Ultimate simulator is a fully realistic road vehicle driving simulator for industrial use (Figure 6.11). The motion system of the vehicle cabin is a hydraulic hexapod combined with an additional linear axis. The compact and lightweight mechanical structure consists of a generic car cockpit, a 200° panoramic on-board screen and driver-access systems, mounted on a novel extended motion system that enables full simulation of acceleration, such as violent changes of direction.

---

[11] http://www.inmotionsimulation.com

Chapter 7

Functional and technical characteristics of VR headsets

Olivier Hugues

7.1 INTRODUCTION

In the previous chapter, we have introduced many technical aspects regarding the use of VR headsets. In this chapter, we are going to present a state of the art of some of the best known headsets that are already on the market or will be soon available. We will detail all the main characteristics of these devices, the software and the hardware when possible, to let the reader know their advantages and limitations. Thereby, the reader of this book may choose the best device to meet the needs of the application he wants to create. VR headsets may have many different uses and it seems very interesting to group them depending on their main functionality. For this, we relied on the Virtual Behavioural Primitives (VBPs) detailed in Chapter 2 by using the following classification: devices which are only used to observe the virtual environment while some others may also be used to move and act in this virtual environment.

7.2 MAIN FEATURES

In the following sections, we are going to present the main characteristics of some of the best known VR headsets to compare them with each other. But it seems also interesting to start by comparing their main characteristics in term of FOV (field of view) and resolution to the psychophysical capabilities of the human visual system.

The list of devices provided at the end of this chapter shows that on average (see Figure 7.1), pixel density and field of view are still far below the capabilities of the human visual system. Indeed, as we mentioned in Chapter 3, the horizontal field of view on average of a person with no visual impairment is approximately 210° horizontally and 140° vertically. This indicates that the quality of most of the VR headsets currently on the market is very poor compared with the main characteristics of the human vision. It is hoped that new technological developments will achieve higher values. It will be easy to increase the pixel density, but improving the field of view is a very hard task with optical issues on which manufacturers have worked for over 20 years.

The pixel density is a very important criterion. Expressed in number of pixels per degree (ppd), it reflects the resolution of the device, regardless of its field of view. This density must be compared to the visual acuity of the human, which is about one minute

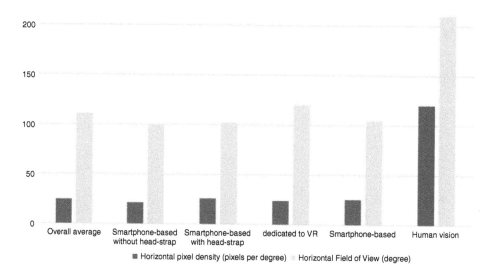

Figure 7.1 Pixel density on average and field of view on average for all the models in the list compared to the psychophysics capabilities of the human visual system.

of angle. We can double this value to obtain an acceptable quality even for users with an acuity beyond the average. The horizontal pixel density of a headset is calculated by the following formula:

– Def_H: horizontal resolution for on eye, in pixel;
– FOV_H: full horizontal field of view, in degree;
– R_H: field of view stereo overlap, in degree;

$$D_H = \frac{2Def_H}{FOV_H + R_H}$$

D_H is often very hard to obtain from manufacturers. When we did not find this value, we have set it to 45° to obtain the approximate pixel density of the models. However, some manufacturers artificially increase the field of view of their devices by designing an optical system with a very limited stereo overlap (R_H). That way, if the horizontal field of view seems to be quite large, the area in which the user is able to perceive 3D stereoscopic images is nil or limited. Figure 7.2 compares the horizontal field of view and the pixel density of devices selected in our list with the optimal characteristics of a "perfect VR headset". To keep things simple, we compare the horizontal field of view, the most important, but it could also be done with the vertical field of view.

One can notice that the pixel density offered by most of the inventoried devices is nearly 25 (ppd) while the visual system would require about 120 ppd. We also note that the largest proportion of the devices provides a field of view from 100° to 120° which is still low in comparison with the human visual system.

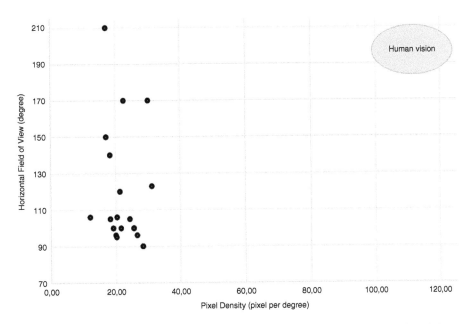

Figure 7.2 Distribution of VR headsets according to their horizontal field of view and pixel density.

7.3 TECHNICAL CHARACTERISTICS OF VR HEADSETS

7.3.1 Smartphone-based headsets

7.3.1.1 Generalities

The performance of this kind of systems depends on the smartphone used which making it difficult to compare them with each other. Indeed, basic hardware specifications such as the definition, the field of vision, the latency, the weight or the contrast depend on the smartphone used. The software also has an impact on the device performance.

These kind of VR headset generally offer a fairly low horizontal field of view around 100°. Even if smartphones look relatively the same regardless of the brand, all the VR headsets provided aren't compatible with all the smartphones on the market. Manufacturers, however, make the effort to design headsets that are compatible or at least partially compatible with several smartphones. Tracking the user's head orientation is done using onboard sensors of the smartphone. They can drift over time and cause discomfort (e.g. if the horizon is not entirely horizontally relative to the physical environment).

In this category, there are two types of VR headsets. Indeed, it is possible to categorize the devices that cannot be fixed on the user's head and must be maintained with the hands, of those which can be fixed on the user's head. The difference is important because the devices belonging to the first category monopolize the hands of the user, which prevents the user from interacting with the virtual environment.

For each smartphone-based device in our list, we consider the use of the compatible smartphone with the highest display resolution. Most of the time, these smartphones are equipped with a 2560 × 1440 resolution display.

Figure 7.3 Google Cardboard (courtesy of Google, Inc.).

Note: In this chapter, information is presented for illustrative purposes only, without any warranty of accuracy, as it's subject to change depending on product evolution and new versions.

7.3.1.2 VR headsets without head band

This is the most basic VR headset one can find nowadays. Composed of one frame, with a very simple design (less than 100 g), it is only useful for keeping up a rudimentary optical system composed of two lenses (one for each eye). The user must keep the device on his face with his hands (one is sufficient). This kind of device has no setting to suit the morphology of the user, it can hurt the user and let the daylight pass which is not so comfortable for the experience.

Many different models exist, but we have chosen to present two of them. One of the best known is certainly the Google Cardboard. The Google Company had sold 5 million units in only 18 months. The dedicated application has been downloaded about 25 million times, with almost 10 million downloads between October and December 2015.

The product proposed by the Google Company (Figure 7.3) was presented to the public in June 2014 at a conference organized by Google and dedicated to developers. It is distributed freely in the form of a cardboard pattern. With a very simple design, without adjustment for the user, this is an far less expensive model in comparison to most of the VR headsets available on the market. It's possible to get it for a few dollars and it's mainly dedicated to marketing purposes or to developers who can use it to test their applications. The company provides an application to test the cardboard, and an API allowing developers to make their own applications compatible. Only the rotation

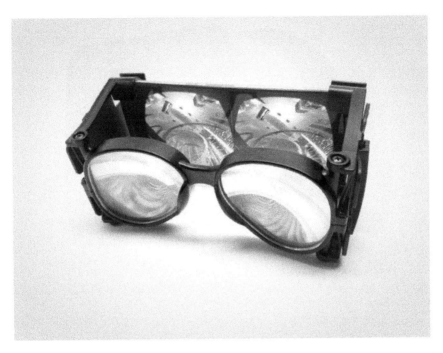

Figure 7.4 Wearality Sky (courtesy of Wearality Corporation. All rights reserved).

of the head is tracked using the sensors embedded in the smartphone, and with a hand used to hold the cardboard in place, it allows only to observe the virtual environment.

The product proposed by Wearality (Figure 7.4) is also smartphone-based, like the Google Cardboard. However, it provides a field of view of approximatively 150°, which is quite large for the field of view of this category. This is done thanks to a Fresnel lens. However, this large field of view reduces the pixel density since the number of pixels remains the same. Indeed, its pixel density is below the average with a value of 13.1 ppd in horizontal and 15.6 pdd in vertical.

Google cardboard applications and games can be used with this device since it is based on the same principle which aims to use sensors embedded in the smartphone.

7.3.1.3 VR headset with head band

These headsets have a very different design than the first two models we have seen above and they are generally more robust. They have a stronger frame, which is used to keep the daylight out, with foam insert to better fit on the face of the user.

They are also smartphone-based but some of them offer the user the ability to adjust one or more parameters such as intraocular distance or sharpness. They are relatively light (less than 350 g), but bigger than previous devices, and most often held on the user's face with an adjustable headband of a few centimeters wide. The headset fits comfortably with a certain degree of stability to support the efforts due to the eccentric position of the smartphone in front of the eyes. Leaving the hands

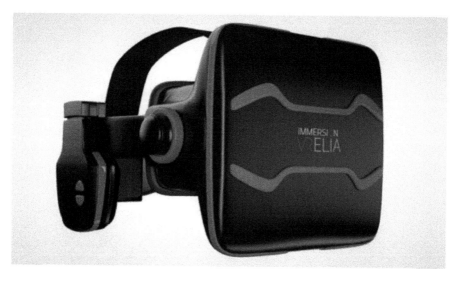

Figure 7.5 BlueSky Pro (courtesy of VRELIA Immersion).

free and allowing to use, for example, a gamepad connected to the smartphone with a compatible application, these devices allow the user to move in virtual environments and sometimes interact with different virtual objects.

This kind of device is intended to be used longer than the last category and, because they fit better, it becomes necessary to have an air circulation mechanism, otherwise, the steam may obstruct the vision of the user, and the heat generated by the smartphone (which comes from the processor, the graphics card and the battery) can be a source of discomfort. There is often a compromise between the circulation of the air and the parts dedicated to block the daylight.

This kind of headset provides, on average, a field of view slightly higher than 100° and a pixel density close to 16 pdd. We will present in the following pages the main devices in this category. However, it's not possible to draw up the list of all the devices in this category because their numbers exploded in the last several months. Furthermore, the number of projects dedicated to design and develop this type of device literally exploded between 2014 and 2016 via a large number of companies. It is necessary to note that a number of these projects was not finalized, and some were stopped, either due to the lack of funding, or because technical issues were not resolved.

The product proposed by the company VRELIA Immersion (Figure 7.5) is based on the use of a smartphone, and thanks to its fairly large frame, it can hold two screens with a 1080 × 1920 display. It is equipped with an audio headset and the frame is both flexible and rigid.

The field of view is above the average with 123°, and despite a large number of horizontal pixels (up to 3840), it has a pixel density in the average (17.9 ppd). Like many other VR headsets, different settings are available to adjust the focus and the distance between the two lenses.

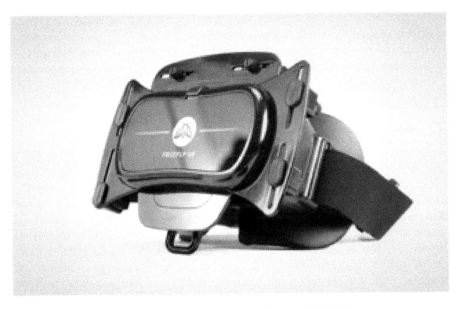

Figure 7.6 Freefly VR (courtesy of Proteus VR Labs).

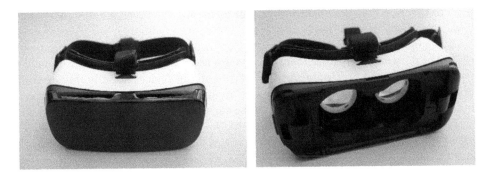

Figure 7.7 Gear VR.

This device is only able to track the rotations of the user's head, and it is compatible with all the "google cardboard" applications. The user is also able to use wireless joysticks to move in the virtual environment.

The product designed by the company Proteus VR Labs (Figure 7.6) proposes a quite large horizontal field of view of 120°. It also provides horizontal and vertical controls to properly maintain the smartphone in front of the eyes of the user. It has a ventilation system to limit the steam and two lenses of about 42 mm diameter. If we consider the best definition of smartphone compatible with this model, the pixel density is about 16 ppd.

The product designed by Samsung (Figure 7.7) is certainly one of the most used smartphone-based headsets even though the number of sales is kept secret and although the horizontal field of view is fairly low with 96°. But thanks to this limited field of view and to the use of a smartphone with a resolution of 2560 pixels, it provides a pixel density of 20 ppd, which is above average.

It weighs 379 g, it will remain stable on the head but only the user's head rotation is tracked using the smartphone's onboard.

7.3.2 Virtual Reality headsets

7.3.2.1 Generalities

These headsets are profoundly different in term of hardware than the devices previously presented. Indeed, they have dedicated electronics rather than using a smartphone, which is specifically designed to meet the needs of virtual reality applications and provide better performances. The onboard inertial unit and/or infrared cameras can detect rotations of the head much faster than smartphones do, with low drift, which limits the time needed for a user movement to be fully reflected on the display screen (motion-to-photon latency). Moreover, some of these devices are equipped with a system to measure the linear motion of the head in the three directions. However, while the angular rotations are taken into account in 360° on all three axes, it turns out that the tracking area may vary from a few centimeters to a few meters depending on the technology used (see Chapter 5). The computing power requires heavy-board electronics. This has a significant impact on the device's weight, and also on energy demands, which makes the use of cables compulsory for now and restrict the movements of the user. It is possible to hope that technological advances allow the use of wireless solutions in a few years.

While most of these devices in this category provide some of the same features, it is however possible to separate them into classes according to their horizontal field of view. On average, the horizontal field of view of all the devices listed in this section is about 120° and the vertical field of view is about 100°. The average pixel density is about 16 ppd, which remains well below the capability of the human visual system.

Later in this chapter, we propose an overview of some recent devices dedicated to virtual reality according to their field of view and if they have an integrated system to track the user's gaze direction.

7.3.2.2 With median field of view

Most of the current devices are in this category but we cannot present all of them here. We present in the following pages three of the most popular devices which have been designed by the most active companies during the last few months.

The Rift (Figure 7.8) was released in early 2016 by the Oculus Company, after the development of two previous versions (DK1 and DK2). The Rift come with OLED displays, each of which offers 1080 × 1200-pixel resolution for each eye. These displays bring the final pixel resolution to 2160 × 1200. With a horizontal field of view of about 106° (full stereoscopic display), and a pixel density slightly greater than 15 ppd, its optical system isn't the best on the market. However, it has a high frame rate for motion tracking data (1000 Hz) and a screen refresh rate of 90 Hz. In doing so, it is possible to expect a relatively low latency, but no value is provided by the manufacturer, despite the efforts that have been made by the company on this aspect. But

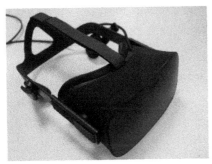

Figure 7.8 Rift of Oculus.

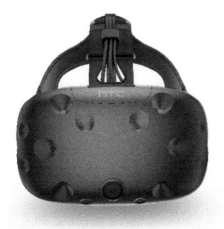

Figure 7.9 HTC Vive (courtesy of HTC & Valve).

one notes that the motion-to-photo latency isn't only described by these data and also depends on the application itself. This device has a full 6 degrees of freedom rotational (yaw, pitch, roll) and translational tracking, which has to be installed next to the user and still requires the use of a cable for the data. Nevertheless, it has an integrated 3D audio system that was the subject of several patents and it has many settings to adjust the brightness and contrast of the screen, or to set the distance between the display and the eyes of the user. It is sold with two controllers allowing the user to manipulate objects in the scene, or to move through the virtual environment. The company also released a Software Development Kit (SDK) that allows developers to make applications compatible with this device.

The HTC Vive (Figure 7.9), the latest product from HTC and Valve Corporation was the consumer version released mid 2016 after the "HTC REVive" (Vive Edition Developer). The device uses two screens, one per eye, each having a resolution of 1080 × 1200. This headset is designed to utilize "room scale" technology with an

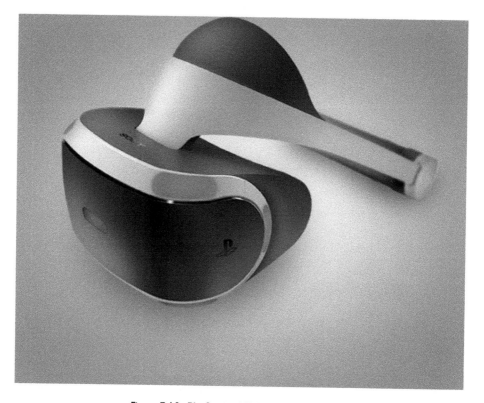

Figure 7.10   PlayStation VR (courtesy of Sony).

"inside-out" tracking. Two lighthouse stations equipped with a laser and a camera should be positioned in the room where the product will be used that tracks the user's movement. Thus, the laser reflects off several diodes positioned on the helmet. This reflection is captured by the cameras and analyzed to compute the orientation and the position of the user relatively to a reference frame defined during the installation of the tracking system. This technical solution requires to arrange the room but movement within a diagonal area of up to 5 m is supported. This version is also equipped with a front-facing camera that blends real world elements into the virtual world to prevent falls or injuries on physical objects. Two handheld controllers are also provided which offer the same functionalities as the Oculus ones.

The screen refresh rate of the PlayStation VR device from the Sony Company, PS VR (Figure 7.10), with 120 Hz, is the same than the Glyph designed by the society Avegant. However, it has a lower resolution with 1920 × 1080 pixels, thus providing a pixel density lower than 14 ppd and a horizontal field of view of 95°, two characteristics which are not in its favour. The manufacturer announced a latency value of 18 ms, which is very satisfying but very hard to confirm. Generally, it is very hard to compare these value between manufacturers because many of them don't communicate this information. The product has also an integrated system capable of detecting the user's rotational movements at 1000 Hz.

Figure 7.11   Claire Full HD (courtesy of VR Union).

It is only compatible with the video game console designed by Sony Company and there is no SDK that allows developers to make applications compatible with this device outside the context of the video game console.

7.3.2.3 With a large field of view

The Claire FullHD (Figure 7.11) designed by the VR Union Company offers one of the largest fields of view available today, with nearly 170°. However, despite the two integrated screens with 1920 pixels horizontally, the large field of view involves a very low pixel density, among the lowest in its category, with only 13 ppd. To solve this problem, the company has upgraded this model, also known as Claire 22 M, with two screens with a total definition of 5210 × 1440 pixels but it's not yet available on the market. Despite this number of pixels, with a density of about 17.2 ppd, it is not yet sufficient to meet the capabilities of the human visual system, but slightly higher than the average of this category.

The StarVR (Figure 7.12) is the product proposed by the Starbreeze Company. This device offers the widest field of view of all the VR headsets available on the market with 210° horizontally and 130° vertically. The company has developed a set of custom Fresnel lenses that make the image appear all around the user's face. Still in development in mid-2016, it is the only device known today able to provide a field of view in accordance with the human visual system. In addition, it has two screens with a very large number of pixel (5120 × 1440). Thanks to this specificity, and despite the fairly large field of view, the pixel density offered by this model remains at the average level, with 17 pdd.

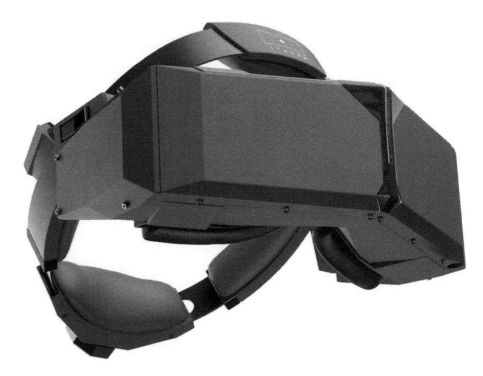

Figure 7.12 StarVR (courtesy of StarBreeze).

7.3.2.4 With eye-tracking system

To track the user's eyes direction, there are two possibilities. On the one hand, tracking is done by upgrading the headset with an eye-tracker module specifically designed to fit in the frame. On the other hand, the headset is able to track the user's eye direction by design with a built-in eye-tracker system.

VR Headsets with added eye-tracker system

Companies that sell eye tracking technology can adapt their products or design new one in a way to be compatible with some existing VR headsets. SensoMotoric Instruments (SMI), for example, are able to upgrade an existing Oculus Rift (not all the hardware versions are available to date) and add eye tracking functionality. It is firstly necessary to buy the headset directly from the Oculus Company, then send it to the SMI Company that will make the upgrade. Then, with the SDK provided by SMI, developers can get the data from the eye tracker at a frequency of 60 Hz, for a horizontal field of view of 80° and a vertical field of view of 60°. The company announced a sensor's accuracy in the range of 0.5° to 1° with a calibration phase done with less than three points.

For its part, Starbreeze Company has partnered with Tobii Company which is a world leader in eye tracking technology to design a built-in eye tracker directly into the StarVR headset.

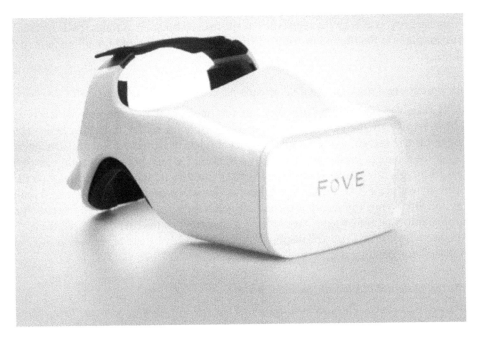

Figure 7.13 Fove (courtesy of Fove Company).

VR Headsets with built-in eye-tracker system

The device designed by the FOVE Company offers a native eye tracker system. If the model is not yet available on the market, it has been founded by a successful crowdfunding campaign in 2015 and the developer kits will be released in fall 2016 according to the company.

The Fove (Figure 7.13) product designed by the Fove Company is the only one headset which is able to natively track the direction of the eyes of the user. With two infrared systems for each eye, the eye tracking sensor seems to be able to track the direction of each eye with an accuracy of 0.2° at a frequency of about 120 Hz. It is also equipped with head tracking sensors that track rotations and translations of the head of the user. It will be equipped with a WQHD display screen (2560 × 1440) and the company announced a field of view of more than 100°.

7.3.3 Augmented Reality headsets

7.3.3.1 Generalities

Although they show slightly different features than other VR headsets, we can also include in this review the headsets that are dedicated to augmented reality and class them in several categories. Whether monocular or binocular, some devices are based on conventional screens (video see-through), and some devices are based on semi-transparent screens (optical see-through).

7.3.3.2 With conventional screen

In this category, one or two screens such as those used for smartphones (LCD or OLED, for example) are positioned in front of the eyes of the user and hide all or part of his field of view.

With partial occultation of the field of view

In this case, the screen hides a portion of the field of view of the user which is able to observe directly the environment with his peripheral vision and when he is taking his eye off the screen. For example, the Jet model, designed by Recon Instruments, consists of a pair of glasses with a slightly deported screen below the right eye of the user. Although this model is equipped with a camera, it only serves to take pictures of the environment (like the Google Glass) and it is not used for an augmented reality functionality. While the screen is turned off when the user does not need to consult the supplied information (to save battery life), it may disturb the user. However, these type of device with classic deported display screen has the advantage of providing sufficient brightness for daytime use.

With complete occultation of the field of view

With these products, the screen hides all of the field of view of the user. This specification makes them very close to the VR headsets we discussed previously. The major difference comes from the presence of one or more digital camera embedded on the headset, which capture the real environment, allowing the user to see what he should have seen if he had no screen in front of his eyes. Without the use of this camera sensors, these devices can be used like a traditional VR headset and the constraints are similar to those encountered with these kind of devices. But efforts have to be done in order to improve the motion-to-photon latency to prevent disorientation and motion sickness which could break the immersive experience. Furthermore, autonomy and power consumption are also a big challenge, since augmented reality is intended for use on the move, unlike virtual reality. This category includes the ImpressionPi device. It has two digital camera sensors (RGB camera and depth camera) to observe the environment. The data captured by the RGB sensor is used to provide an augmented reality functionality by displaying on the screen the environment of the user. The depth camera is used to reconstruct the user's environment in three dimensions. Using this reconstruction, 3D objects (such as texts, images or 3D models) can be integrated properly in the image on the screen by taking into account the occlusions with the real environment.

7.3.3.3 With optical see-through

Several techniques have been proposed in recent years and all these solutions can be grouped into two categories. There are technologies based on curved mirrors and technologies based on light rays (waveguide or light guide). The first technology uses curved mirrors disposed in front of the eyes of the user with an offset in order to not restrict the user field of view. This solution involves large image distortion that must be corrected by embedded electronics or complex optics. These adjustments add complexity to the system, add weight, increase the volume of the device and have a negative impact on the autonomy and the resolution. The model M100 from Vuzix

Figure 7.14 Laster WAVE (courtesy of Laster Technologies).

and many others devices are using this technology. The second technology is based on light guide and this is the most used solution for AR headsets over the past years. With this technology, the user perceives the real environment through a semi-transparent mirror held in front of his eyes. A projected light rays come from a remote source, guided through an optical channel. Thus, light source and electronics may be deported from the projection surface viewed by the user, which facilitate their integration in the headset. Several optical principles are used in this type of device such as diffraction, holography, polarization or reflection.

The device designed by Google, as known as Google Glasses, is an example of a product that uses optical reflection. This is a pair of glasses with a monocular see-through display. This screen is positioned in the upper part of the field of view of the user. A rear-mounted pico-projector produce an image conduced from the source to the projection surface by reflection. The main disadvantage of this type of system is the low brightness and contrast ratio, making it very difficult to use under daylight conditions. In addition, the energy required to operate such a projector is still important and a compromise is necessary between the autonomy of the device and the weight of the battery. Laster Technologies is also involved in this market since 2005 with several models, and more particularly with a monocular see-through display named Laster WAVE (Figure 7.14). With a diagonal field of view of about 25° and a brightness of 4200 cd/cm^2, the system is fully wireless with an autonomy of more 3 hours. The Laster WAVE is completely autonomous and runs with an embedded Android OS. The product has also an embedded head tracking system with 9-axis IMU (accelerometers, gyroscopes, compass), and a built-in camera (in its Premium version).

During 2015 Microsoft has released a new device, called HoloLens, which is a headset composed by a pair of binocular glasses. A lot of efforts have been made on both hardware and software despite the fact that the FOV is approximately 35° and there is no use of light field display in this version. During mid-2016 Microsoft released the design of what it called a "holographic processing unit" (or HPU) which is powered

24 digital signal processor cores, capable of around 1 trillion operations per second to compute the data provided by all the embedded sensors. The product is equipped with many sensors (about a dozen) dedicated to capture the environment of the user and his gestures. It's also equipped with a spatial sound system. Few depth sensors are able to reconstruct a 3D scene of the surrounding environment of the user and this reconstruction is used to improve the merge of real and virtual objects.

During 2016, Intel announced a new project called "Alloy". The device is wireless and doesn't need any smartphone. As the Microsoft Hololens, this device has his own embedded electronics components with a battery and two RealSens™ sensors which are dedicated to sense depth and track user motion.

We also can note that Google is supposed to release Daydream during the 2016 fall season with a higher quality device than the Cardboard to compete with the GearVR. Daydream will be optimized for Android, and accompanied by a remote controller that let the user wirelessly interact and play with the virtual world. The main advantage of this device is his compatibility with a large amount of phone makers such as Samsung, HTC, LG, Huawei, Xiaomi, Asus, ZTE and Alcatel.

"The Void" is another project that is similar to a "laser tag": a dedicated big space where several participants can experience an adventure in virtual reality. It will take place in a hangar with protective foam on the walls. The players will use a VR headset designed specifically for this purpose with a field of view about 160°, data gloves and haptic gaming vest.

The chip maker Qualcomm has revealed a reference platform for a standalone virtual reality headset, which it's calling the Snapdragon VR820. The VR820 was created in partnership with Chinese electronics company GoerTek, built on Qualcomm's Snapdragon 820 mobile processor and software development kit. It is designed to help developers create the hardware and the software to create immersive VR experiences. It will be equipped with two cameras for eye tracking, a feature that was also announced in the Fove headset and in the StarVR device.

One note that there is a great number of devices available but China is one of the leading country. Indeed, Alibaba and Taobabo have sold several hundreds of thousands of smartphone-based headsets. There are more than one hundred different devices available on the Chinese market, mainly low-end products, similar to the Google Cardboard. Currently, the marked is dominated by 3Glasses, DeePoon and Baofeng Mojing with 1 million units sold by Baofeng during the first quarter of 2016.

7.4 CONCLUSION

In this chapter, we have proposed an overview of VR headsets that can be used for 360-video, virtual reality and/or augmented reality. We first note that the number of devices already available on the market is quite large and there is a lot of different types of devices with heterogeneous technical solutions. All of these devices have advantages and disadvantages that is necessary to consider in order to properly select the device depending on the application. It should be noted that at the time of writing this book, there is no device that can be compared to the capacity of the human visual system. Many technological improvements are still needed to achieve this and many years of research and developments will probably be required to reach this goal.

Summary of VR headsets specifications

Name	Maker	Type	Screen Type	Size (pouce) min	Size max	Res. (px) H	V	Res. (ppi)	Res./eye (px) H	V	Pixel density (ppd) H	V	Refresh rate (Hz)	FOV(°) 1 eye D	H	V	2 eye D	H	V	Head R	T	Eye	Weight (g)	DIO (mm) Settings	min	max	Web site
Bluesky Pro	Immersion Vrelia	With Fixation	–	–	–	3840	1080	–	1920	1080	22,86	–	–	–	–	–	–	123	–	Y	–	N	–	–	–	–	http://immersionvrelia.com/bluesky
AirVR	Vrelia AirVR I	With Fixation	–	–	–	1920	1080	–	960	1080	13,24	–	–	–	–	–	–	100	–	Y	–	N	–	–	–	–	http://getairvr.com
Freefly VR	Proteus VR Labs	With Fixation	–	4,7	6,1	2560	1440	–	1280	1440	15,52	–	–	–	–	–	–	120	–	Y	–	N	–	–	–	–	https://www.freeflyvr.com/
Dlodlo H One	H One	With Fixation	–	–	–	2560	1440	–	1280	1440	15,52	–	–	–	–	–	–	120	–	Y	–	N	–	–	–	–	http://www.dlodlo.com/en/
Cmoar	Cmoar VR	With Fixation	–	–	–	2560	1440	–	1280	1440	17,07	–	–	–	–	–	–	105	–	Y	–	N	–	–	–	–	http://cmoar.com
Homido	Homido	With Fixation	–	4	5,7	2560	1440	–	1280	1440	17,66	–	–	–	–	–	–	100	–	Y	–	N	–	–	–	–	http://www.homido.com
VR One	Zeiss	With Fixation	–	4,7	5,2	2560	1440	–	1280	1440	17,66	–	–	–	–	–	–	100	–	Y	–	N	–	–	53	77	http://zeissvrone.tumblr.com
Durovis Dive 5	Shoogee	With Fixation	–	2	5	2560	1440	–	1280	1440	17,66	–	–	–	–	–	–	100	–	Y	–	N	150	–	–	–	https://www.durovis.com/product.html?id=1
Durovis Dive 7	Shoogee	With Fixation	–	–	7	2560	1440	–	1280	1440	17,66	–	–	–	–	–	–	100	–	Y	–	N	–	–	–	–	https://www.durovis.com/product.html?id=1
Visus VR	Visus	With Fixation	–	–	–	2560	1440	–	1280	1440	17,66	–	–	–	–	–	–	100	–	Y	–	N	–	–	–	–	http://www.visusvr.com
Xingear XHVR	–	With Fixation	–	–	–	2560	1440	–	1280	1440	17,66	–	–	–	–	–	–	100	–	Y	–	N	–	–	–	–	http://www.imcardboard.com/xingear.html
Pinch VR	Pinch VR	With Fixation	–	–	–	2560	1440	–	1280	1440	17,66	–	–	–	–	–	–	100	–	Y	–	N	–	–	–	–	http://hellopinc.com
Dior Eyes	Dior	With Fixation	–	–	–	2560	1440	–	1280	1440	17,66	–	–	–	–	–	–	100	–	Y	–	N	–	–	–	–	http://digitaslabsparis.com/post/119609782829/dior=AS1eyes
Gear VR (Note 4)	Samsung/ Oculus	With Fixation	Super AMOLED	5,7	–	2560	1440	515	1280	1440	18,16	–	60	–	–	–	–	96	–	Y	–	N	379	–	55	71	http://www.samsung.com/global/microsite/gearvr/gearvr_specs.html
Gear VR (S6/Edge)	Samsung/ Oculus VR	With Fixation	AMOLED OLED	5,1	–	2560	1440	576	1280	1440	18,96	–	60	–	–	–	–	90	–	Y	–	N	350	–	55	71	http://www.samsung.com/global/microsite/gearvr/gearvr_specs.html
Impression Pi	–	With Fixation	–	–	–	2560	1440	–	1280	1440	18,96	–	–	–	–	–	–	90	–	Y	Y	N	–	–	–	–	http://www.impressionpi.com

(continued)

Summary of VR headsets specifications (Continued).

Name	Maker	Type	Screen Type	Size min (pouce)	Size max	Res. H (px)	Res. V	Res. (ppi)	Res./eye H (px)	Res./eye V	Pixel density H (ppd)	PD V	Refresh rate (Hz)	FOV 1 eye D (°)	1 eye H	1 eye V	FOV 2 eye D	2 eye H	2 eye V	Track Head R	Track Head T	Track Eye	Weight (g)	DIO Settings (mm)	DIO min	DIO max	Web site
Evomade	Viewbox	With Fixation	–	–	–	2560	1440	–	1280	1440	18,96	–	–	–	–	–	–	90	–	Y	N	N	–	–	–	–	http://evomade.com
Claire FullHD	VR Union	For VR	–	–	–	3840	1080	–	1920	1080	17,86	–	–	–	–	–	–	170	–	Y	N	N	–	–	–	–	http://vrunion.com
Oculus Rift DK1	Oculus VR	For VR	LCD	7	–	1280	800	216	640	800	8,48	10,78	60	–	–	–	110	106	95	Y	Y	N	380	Adjust IPD, Adjust focus, 2 ind. screen lenses	–	–	https://www.oculus.com/en=ASIus/
Oculus Rift DK2	Oculus VR	For VR	OLED	5,7	–	1920	1080	386	960	1080	13,71	15,21	75	–	–	–	100	95	95	Y	Y	N	440	Adjust IPD, Adjust focus, 2 ind. screen lenses	–	–	https://www.oculus.com/en=ASIus/
Gameface	Gameface Labs	For VR	OLED	–	–	2560	1440	–	1280	1440	13,84	–	75	–	–	–	–	140	–	Y	N	N	–	–	–	–	http://gamefacelabs.com
StarVR	Starbreeze	For VR	LCD	5,5 × 2	–	5120	1440	–	2560	1440	17,07	22,59	60-90	–	–	–	247	210	130	Y	Y	N	–	Adjust IPD	0	–	http://starvr.com
Playstation VR	Sony	For VR	OLED	5,7	–	1920	1080	386	960	1080	13,71	–	120	–	–	–	100	95	–	Y	Y	N	–	–	–	–	https://www.playstation.com/en=ASIus/explore/playstation=ASIvr/
OSVR HDK 1.3	Razer/Sensics	For VR	OLED	5,5	–	1920	1080	401	1080	960	14,90	–	60	–	–	–	–	100	–	Y	N	N	–	Adjust focus	–	–	http://www.razerzone.com/osvr
The Rift	Oculus VR	For VR	OLED	–	–	2160	1200	456	1200	1080	15,48	16,99	90	–	–	–	110	95	–	Y	Y	N	470	Adjust IPD, Adjust focus	–	–	https://www.oculus.com/en=ASIus/
HTC Vive	HTC/Valve	For VR	OLED	–	–	2160	1200	–	1200	1080	15,89	–	90	–	–	–	110	106	–	Y	Y	N	555	Adjust focus	–	–	http://www.htcvive.com/us/
Totem	VRVANA	For VR	OLED	–	–	1920	1080	–	960	1080	12,80	–	120	–	–	–	110	105	–	Y	N	N	400	Adjust focus	–	–	http://www.vrvana.com/fr
FOVE	FOVE	For VR	OLED	5,8	–	2560	1440	506	1280	1440	17,66	–	60-90	–	–	–	100	100	–	Y	N	Y	400	Adjust focus	–	–	http://www.getfove.com
Glyph	Avegant	For VR	Micro-projection	–	–	2560	1440	–	1280	720	28,44	–	120	–	–	–	–	45	–	Y	N	N	850	Adjust IPD	–	–	https://avegant.com
Claire 22M	VR Union	For VR	OLED	–	–	5120	1440	–	2560	1440	23,81	–	–	–	–	–	–	170	–	–	–	–	–	–	–	–	–
Wearality Sky	Wearality	without Fixation	–	–	–	2560	1440	–	1280	1440	13,13	15,66	–	156	130	123	165,5	150	123	Y	N	N	–	–	–	–	http://www.wearality.com
Cardboard	Google	without Fixation	–	5,2	–	2560	1440	–	1280	1440	17,66	–	–	–	–	–	–	100	–	Y	N	N	90	–	–	–	https://www.google.com/get/cardboard/

Chapter 8

Comfort and health

8.1 COMFORT AND HEALTH ISSUES

In the real world, man builds a coherent representation based on all the sensory stimuli received. In the virtual world, users will therefore seek to interpret coherently what they perceive based on their experience in the real world, despite the sensorimotor discrepancies:

– Between several senses (for example locomotion on a treadmill resulting in a discrepancy between the vision and vestibular systems, Figure 8.1);
– Between the senses and motor responses (when manipulating virtual objects without force feedback, meaning users have no sensation of their weight, for instance).

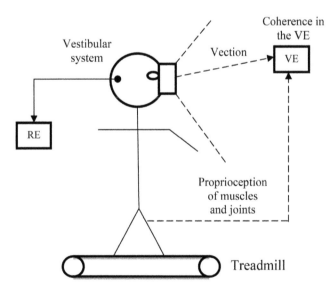

Figure 8.1 With movements on a treadmill, there is coherency between the Virtual Environment (VE) vision and the proprioceptive organs in muscles, tendons and joints; however, vestibular systems are incorrectly stimulated because the person is standing still in the Real Environment (RE).

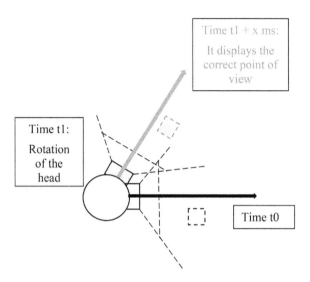

Figure 8.2 The latency time is the time lag between the user's action and the response perceived, as a result of his action.

It is currently hard to determine whether the subject immersed in a given virtual environment is able to perceive and make sense of these environments effortlessly, by managing to overcome such sensorimotor discrepancies. In virtual environments, another potential discomfort for users is the slowness of the reactions in the virtual environment in relation to the actions they have taken. This latency lag, which depends on the technological performance of the VR device, will therefore affect the behaviour of users. It can be described as a sensorimotor discrepancy, but one with a temporal criterion. This latency results from adding several response times together: the motor interface (joystick, force feedback interface, etc.), the computer processing of the control signal received, the time it takes the computer to create the response required (image rendering, etc.) and the transfer of the response to the sensory interface (in our basic case, the time required to display the correct image in the VR headset) (Figure 8.2).

In VR applications, not all of our senses intervene to create the desired simulation. However, some are to be taken into account for the sensorimotor discrepancies induced by VR techniques, such as:

- Vision;
- Hearing;
- Skin sensitivity;
- Kinaesthetic sense (vestibular systems, sensory receptors in the joints, etc.);
- Muscle proprioception.

An awareness of movement is given by the position of the different parts of the body and their relative mobility, as well as by the force of muscular contractions during the movement. In addition to the proprioceptive organs of the muscles, tendons and joints, the organs that provide the sensation of motion are those located in the receptors of both vestibular systems in the inner ears as well as, to a large extent, the vision (see chapter 3 for detailed presentation of the human senses). Independently of the technical solutions used, and even when using a perfect VR headset – i.e. one that offers the same visual quality as that of the real world – a discrepancy will always be present for certain types of applications. The aim is not merely to resolve technological issues in the near future to eliminate all such discrepancies and allow users to perform any sensorimotor activity in a virtual environment, as some limitations will always be encountered. In principle, we are creating a new artificial world with its rules, limitations and potentialities, some of which will exceed or differ from those encountered in the real world. Based on experience and some classic and well-analysed cases we are well aware that people are able to overcome some discrepancies by adjusting to them. And some adjustments occur almost naturally: the most explicit example is the virtual motion in front of a small computer screen or games console. Despite this discrepancy between visual-vestibular cues, users will still perceive the Real Environment (RE)[1] in their peripheral vision and to some extent will experience the sensation of their own movement, if the Virtual Environment (VE) displays moving images. The phenomenon of vection is responsible for this sensation of movement, despite the fact that the vestibular systems of the person who is motionless in the real world are not detecting any movements. It is extremely rare to find people who do not adjust to this type of virtual movement in front of a screen that does not cover their peripheral vision. For observations via a VR headset, the problem is much more complex.

The use of virtual reality headsets can trigger a number of health and comfort issues, which are mainly caused by certain categories of virtual reality applications, and make users feel unwell primarily as a result of:

- Psychological activity of the disturbed subject in the virtual environment;
- Poor interfacing between the subject's visual system and the VR headset;
- Unsafe technological devices;
- Sensorimotor discrepancies:

8.2   INTRODUCTION TO SENSORIMOTOR DISCREPANCIES

It is not only the visual-vestibular discrepancy that is involved, although this particular discrepancy is the hardest to manage. People using VR applications face a large number of sensorimotor discrepancies. It is therefore worth classifying these sensorimotor discrepancies and analysing them to obtain more than just a mere inventory of all of the recommendations VR designers should follow to help them improve the well-being of users. We will first analyse this problem from a VR interfacing approach (based on the $3I^2$ model). The third level, "functional Immersion and Interaction", concerns the activities of users in a VE. These can always be divided into a few basic behaviours,

---

[1] As their peripheral vision is in the real environment, the posture of users in the RE remains stable due to the coherence between their peripheral vision and vestibular systems, and they do not experience any sensations of sickness or discomfort.

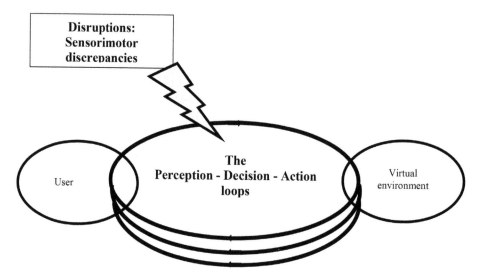

Figure 8.3 Sensorimotor discrepancies disrupt the level of sensorimotor I^2.

called "Virtual Behavioural Primitives" (VBPs), and grouped into four categories (see chapter 2):

– Observing the virtual world;
– Moving in the virtual world (navigation);
– Interacting with the virtual world, mainly by handling objects;
– Communicating with others or with the VR application.

The first three are difficult to manage at the level of the sensorimotor discrepancies, and we will only analyse in detail these three. The last of these VBPs also poses a few specific problems. In addition, and whenever feasible, the metrological characteristics of the behavioural interfaces must offer an extremely high performance in order to match the psychophysical capacities of the senses and/or the motor responses of man as closely as possible. In practice, the quality of the sensory stimuli and motor responses must be just as satisfactory as those of the VBPs required by the application.

It should be stated that the main difficulty from immersing and interacting with subjects in a virtual environment stems from **disruptions** to the "Perception, Decision, Action" (PDA) loop: the subject Perceives his environment and Decides to Act receiving perceptive feedback from the actions taken. Technological devices inserted into the VR loop interfere with its operation. In practice, there are several PDA loops when looking at all the senses separately. We will explain below how a careful selection of the Virtual Behavioural Primitives, interfacing devices, cognitive processes and Behavioural Software Aids can control these disruptions to allow subjects to act effectively in a VE, via artificial PDA loops. However, we will first detail the disruptions caused by sensorimotor discrepancies, Figure 8.3.

Perceptions and actions must be perceived coherently the user's brain. We believe that this should be the main working hypothesis used to analyse the impact of all these types of sensorimotor discrepancies. We can also make other hypotheses, rated according to the following order of importance:

H1: the brain seeks to **perceive a coherent world** using all of the sensory stimuli received and from the motor responses in the "Perception, Decision, Action" loop in the VE, as it has done in the RE since childhood, when developing sensorimotor intelligence.

H2: the brain **anticipates its motor responses** based on the sensory stimuli received.

H3: **the PDA loop is disrupted** in all virtual reality applications.

H4: a subject's activities in a virtual environment **almost always create** sensory and sensorimotor **discrepancies.**

H5: VR techniques **disrupt the instantaneousness** of the PDA loop as a result of the latency lag, which, for technical reasons, is always present.

H6: **vision is a predominant sense** as far as sensory and sensorimotor discrepancies are concerned.

H7: when the brain is unable to make its perceptions and actions coherent in a virtual environment, this will result in, **at the very least, discomfort or even a sensation of feeling unwell and, in the worst-case scenario, rejection** by users.

H8: users can **adapt** to these discrepancies in a VE through learning.

H9: users have a **wide and varied range of coping skills.**

H10: for the vast majority of the population, some discrepancies are known and can be **beneficial and exploited**, while others will be **disruptive** to varying degrees.

8.3 TAXONOMY OF SENSORIMOTOR DISCREPANCIES

8.3.1 Introduction

While most discrepancies can be troublesome – sometimes even totally unacceptable – they can also have a **beneficial** effect when the subject, by seeking to understand the coherence of the world, "believes in the visual simulation". This is reminiscent of the classic case, outlined above, of visual-vestibular discrepancy with peripheral vision in a RE: that is, users believe they are moving due to the phenomenon of vection without placing any demands on their vestibular systems. Some discrepancies can, therefore, offer effective solutions. And this same logic can be extended to the use of **pseudo-haptics**[2] (Lécuyer, et al., 2000). The basic experiment consists of applying a variable force through a spring with constant stiffness to the user's hand, while visually displaying a variation in the movement of the object manipulated by the user. Changing visually the virtual object's amplitude of movement by the means of a given force on the spring will result in the tester perceiving (feeling) a different stiffness in the spring. The reason for this is that the tester's brain is striving to make his action coherent in reference to the "internal (mechanical) model" his brain has acquired over the course

---

[2]Haptic refers to the sense of touch and the muscular proprioception that is stimulated in the VR via the haptic inter faces, which are made up of two categories: tactile interfaces and force feedback interfaces.

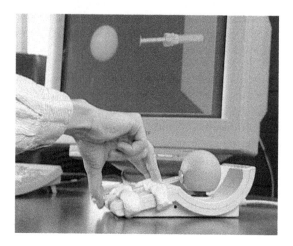

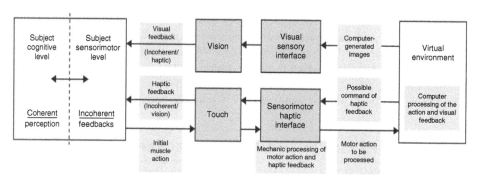

Figure 8.4 Experimental device and diagram of the pseudo-haptic principle.

of past experiences. This phenomenon, which is used in virtual reality applications, generates haptic sensations by modifying the visual feedback when the user acts in the virtual environment, **without needing an active haptic feedback interface** controlled by the simulation. Figure 8.4 explains this pseudo-haptic feedback and shows that the subject is stimulated by an incoherent series of visual stimuli and real haptic stimuli, as the latter do not vary. Pseudo-haptic feedback would seem to correspond to a reinterpretation of these stimuli and an optimal visual-haptic perception of a world that should remain or become coherent for the subject.

In the two previous cases, vision is the predominant sense for the user's perception that is "users will believe what they see" (hypothesis H6).

> *These discrepancies are described as beneficial when, as long as they do not cause discomfort and make people feel unwell, they make the simulation appear more realistic on a multi-sensory perceptual level, providing a closer reflection of what happens in the real environment, by failing to properly stimulate one of the senses.*

Conversely, discrepancies are described as disruptive when they imply discomfort for users or make them feel unwell.

Some are very disruptive, meaning that the majority of users will be unable to tolerate them. Designers of VR applications are strongly recommended to never rotate the VE in the VR headset without the user – who is motionless in the RE – having to turn his head, that is without the user having to do the action himself. The reason for this is that the user is unable to *anticipate* his visual perception in the VE, given that his brain has not received information from his own voluntary actions. People attempting to do this test will almost always feel unwell. A fairly similar experience in the real world is to imagine you are a passenger in vehicle instead of being the driver. Anticipating rotations and accelerations is much harder when you are passive in relation to them, despite the vestibular system detecting such movements. People are therefore more prone to motion sickness when they are passengers rather than the driver. Consequently, designers are strongly advised to never design actions that are not controlled by the head (or the hands or the whole body) of the person wearing the VR headset.

A non-exhaustive list of beneficial sensorimotor discrepancies, established on the basis of the three Virtual Behavioural Primitives (VBPs) (observing, navigating and manipulating) is given:

8.3.2 List of 5 beneficial sensorimotor discrepancies

– Observational VBPs:

* *Visual-acoustic discrepancy*: The sound source of a virtual object will be located by its position displayed in the VR headset, even if the sound source is not located to the same visual position of virtual object. It is the "ventriloquist illusion" which creates the illusion that the voice emerges from the visibly moving mouth of the puppet. This well-known illusion exemplifies a basic principle of how auditory and visual information is integrated in the brain to form a unified multimodal percept. When auditory and visual stimuli occur simultaneously at different locations, the more spatially precise visual information dominates the perceived location of the multimodal event. This illusion does not function if there is too much difference between the visual location and the sound location;
* *Oculomotor discrepancy*: stereoscopic vision enhances the three-dimensional perception of objects, even when images are displayed on the flat surface of the screen. The stereoscopic vision enables the illusion of seeing objects in front of or behind the screen! This is achieved due to the convergence of the observer's eyes, despite the fact that there is an "accommodation-vergence" discrepancy and although the accommodation indicates to the brain that the images of the objects are at the level of the screen (Figure 8.5). However, this illusion is only possible within certain limits of retinal disparities (see the "disruptive oculomotor discrepancies")

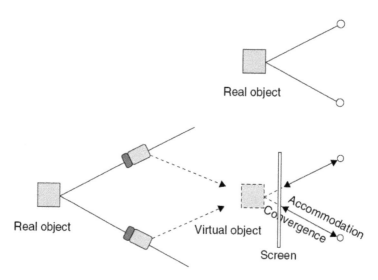

Figure 8.5 With real or virtual cameras (synthesised stereoscopic images), observers accommodate the images on the screen and their eyes converge at a different distance, according to the horizontal parallax of the images.

- Navigational VBP:

  • *Visual-vestibular discrepancy:* this is the classic case of virtual movement by vection without the observer actually moving in the RE. This discrepancy is beneficial because the user accepts as true the movement. The visual-vestibular discrepancy is disturbing to the user as soon as it exceeds certain cinematic movement limits (see the "disruptive visual-vestibular discrepancies")

- Manipulation VBPs:

  • *Visual-haptic discrepancy:* it is the pseudo-haptic which is used in virtual reality applications and generates haptic sensations by modifying the visual feedback when the user acts in the virtual environment, without needing an active haptic feedback interface;
  • *Visual-tactile discrepancy:* it is an interaction technique to simulate textures in desktop applications without a tactile feedback interface. The technique consists in modifying the motion of the cursor on the computer screen, namely the Control/Display ratio. Assuming that the image displayed on the screen corresponds to a top view of the texture, an acceleration (or deceleration) of the cursor indicates a negative (or positive) slope of the texture. Experimental evaluations showed that participants could success- fully identify macroscopic textures such as bumps and holes, by simply using the variations of the motion of the cursor (Lecuyer, Burkhardt, *et al.*, 2004).

The 5 discrepancies are beneficial because the vision is predominant: The subject, by seeking to understand the coherence of the world, "believes in the visual simulation".

But most discrepancies can be troublesome and sometimes even totally unacceptable. A non-exhaustive list of disruptive discrepancies, established on the basis of the three Virtual Behavioural Primitives (VBPs) (observing, navigating and manipulating), is given:

8.3.3 List of 11 disruptive sensorimotor discrepancies

– Observational VBPs:

- *Temporal visual-motor discrepancy*: a specific problem with VR headsets is the latency lag between the observer's head movement and the correct display of the viewpoint corresponding to this movement on the screen of the VR headset. This discrepancy is negligible and imperceptible when it is less than a millisecond, which is not the case with the current VR headsets, unless the headset is fitted with a sensor with mechanical connections linking it to a fixed support in the room and the VR headset. It is worth restating that effective vision requires that the eyes must be stabilised in the observed area. The vestibular system allows the vestibulo-ocular reflex[3] to stabilise the gaze. In a VR application, any lengthy latency lag in between a user's head movement and the effect of this movement displayed on the screen will result in the user's eyes perceiving a delayed movement, which can be out of kilter by a few milliseconds compared with the one perceived by the vestibular system. Anticipations are difficult under such conditions and this causes instability of the gaze (Stanney, 2002);
- *Visual-temporal discrepancy*: there is a disruptive discrepancy if the frequency of images displayed (FPS: *Frames per second*) is too low compared to the needs of the visual system to perceive images without flickering and with continuously moving objects in motion. This is not dependant on retinal remanence (old discounted theory) but on the "phi" phenomenon and the "beta" movement, which are neurophysiologic mechanisms: when the brain perceives, for example, two bright spots alternately displayed at an angular distance close to one another, the brain perceives the first point moving to the position of the second point, which gives an impression of movement on the basis of one-off information. The higher the frequency (FPS of 120, 240 Hz), the more movements will appear fluid;
- *Oculomotor discrepancy*: This is the "accommodation-vergence" conflict which becomes a disruptive constraint if the retinal disparities imposed on the observers exceed a certain limit, which depends on the exposure time, on the visual capabilities of the observer and on other criteria. Some people who

---

[3]The vestibulo-ocular reflex makes the eye swivel in the opposite direction to the head's movement. Vestibular systems, by measuring the head's rotation, order the eyes to move in a direction opposite to the movement of the head to stabilise the retinal images.

are more sensitive to this discrepancy cannot manage to merge or even look at this type of image;

- *Visual-spatial discrepancy*: a disruptive discrepancy is created if the fields of view of the VR headset differ from the fields of view of the camera filming the VE. Some designers use this subterfuge to increase the fields of view for the observer, as those offered by VR headsets are far too narrow;
- *Visual-motor localisation discrepancy*: head movements in the real environment control the visual rotation in the VE, which may differ due to the fact that the rotational sensor is inaccurate or fails to measure the translations. In the latter case, it will not be able to detect any minor head rotations, even if the observer is sitting or standing relatively still.
- *Spatial visual-motor discrepancy*: designers of VR applications may wish to program *an unnatural visual observation*:

  o either amplifying the virtual rotation relative to the rotation of the head, to allow the observer to see over a larger field of vision without having to turn his head too much;

  o or amplifying the virtual translation relative to the translation of the head, to allow the observer to artificially "zoom" his vision;

  o or, and even more disturbingly, the viewpoint does not correspond at all to the viewpoint of the observer's head; for example, the "third-person view" in the VE (or the "objective view"). Observers see their avatar (their own representation) in the VE. Observers can watch themselves or see elsewhere. The viewpoint can also be that of a virtual character such as, for example, a character facing the observer. When this is the case, the observer sees themselves from the eyes of another person. This type of discrepancy, due to an unnatural view, is virtually unacceptable for the majority of the population. However, in practice, nothing prevents some people from adapting to such a view when they have the motivation and take the necessary time. It is worth restating the experiment carried out in the middle of the 20th century by Professor T. Erismann to test a much more complex vision for the brain to process: namely, an inverted vision through a pair of glasses that flipped the view upside down (Figure 8.6). It took around ten days for the wearer to adapt perfectly to this new view, who then had to readjust to normal vision when the glasses were removed;

- *Passive visual-motor discrepancy*: designers of VR applications are strongly advised to never rotate the VE in the VR headset without the user – who motionless in the RE – having to turn his own head. This is an extremely disruptive discrepancy that does not correspond at all to realistic behaviour. Although this type of *unnatural visual observation* may be offered sparingly in some specific applications.

– Navigational VBPs:

- Visual-vestibular (or *visual-proprioceptive*) discrepancy: this is the classic case of virtual movement by vection without the observer actually moving. This discrepancy is well known and described as simulation or motion sickness.

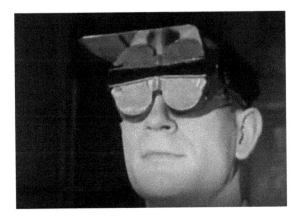

Figure 8.6 An inverted vision through a pair of glasses that flipped the view upside down, experimentation by Professor T. Erismann.

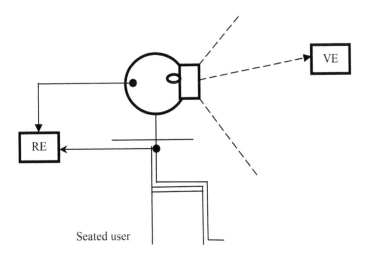

Figure 8.7 As the person is seated, motion in the VE is achieved through vection; however, the vestibular systems are incorrectly stimulated because the person remains motionless in the RE, and this also applies to the proprioceptive organs in the muscles, tendons and joints.

Several theories presented in the appendix at the end of this article explain the discomfort induced by discrepancies of this nature. The visual-vestibular discrepancy is disturbing to the user as soon as it exceeds certain cinematic movement limits (Figure 8.7). The movement command can be manual and operated through the head or the feet of the user; this latter case requires a treadmill. This case is more natural and coherent with vection motion, as coherency exists between the vision and the proprioceptive organs in the muscles, tendons and joints. However, the vestibular systems are poorly

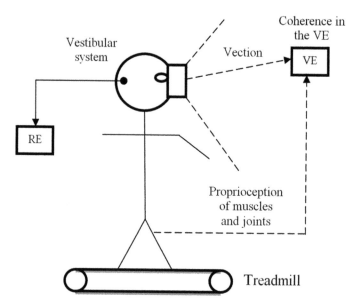

Figure 8.8 With movements on a treadmill, there is coherency between the VE vision and the proprioceptive organs in muscles, tendons and joints; however, vestibular systems are incorrectly stimulated because the person is standing still in the RE.

stimulated, given that the person is standing still in the RE (Figure 8.8). In contrast, with motion simulation interfaces, all of the aforementioned proprioceptive organs and vestibular systems can be properly stimulated. This discrepancy can be eliminated by using an appropriate movement simulation interface although, in such cases, may induce another discrepancy, which is described in the next paragraph;

- *Temporal visual-vestibular discrepancy*: it is possible therefore to correctly stimulate the vestibular systems in relation to movements in the VE by placing the person on a mobile platform. The motion simulation interface can be either a six Degrees-Of-Freedom (DOF) platform or a two to four DOF movable seat, depending on the need for correspondence between real movements and virtual movements created by vection. In this case, the latency lag between the movements of the mobile platform or the seat and the display of the correct point of view should be minimal to avoid any temporal visual-vestibular discrepancies;
- *Visual-postural discrepancy*: when users are standing in a stationary position in RE, they need to control their posture and the vertical position of their body despite any discrepancies. These can occur if the motion in the VE is via vection and the body remains motionless or almost motionless, as the brain is informed of the fixed position of the body by the vestibular systems but also by other proprioceptive stimuli (neuromuscular spindles, Golgi tendon organs and joint receptors);

– Manipulation VBPs:

   • *Visual-manual discrepancy*: when manipulating objects, if the user's real hand is incorrectly positioned compared to the hand he sees in the VR headset (when the hand is actually shown) for technical reasons, this will cause a visual-manual discrepancy and the user will need to adjust to this discrepancy. However, the manipulation can be performed in an unnatural and unreal manner by "tele-operation" (see below).

IN SUMMARY, THE LIST OF 11 DISRUPTIVE DISCREPANCIES

Observational VBPs:

Temporal visual-motor discrepancy
Visual-temporal discrepancy
Oculomotor discrepancy
Visual-spatial discrepancy
Localisation visual-motor discrepancy
Spatial visual-motor discrepancy
Passive visual-motor discrepancy

Navigational VBPs:

Visual-vestibular discrepancy
Temporal visual-vestibular discrepancy
Visual-postural discrepancy

Manipulation VBPs:

Visual manual discrepancy

Disruptive discrepancies all involve the vision, which corresponds to hypothesis H5.

The first five disruptive discrepancies for observational (visual) VBPs are all caused by **technical difficulties** as the VR headsets currently available on the market are not yet "perfect enough": that is, they do not offer sufficiently short latency lags, sufficiently high frame rates, sufficiently broad fields of view and sufficiently precise head tracking movements, as well as stereoscopic screens without oculomotor discrepancy. Obviously, when the **visual observation is unnatural and unreal** (sixth and seventh discrepancies) disruption for the user is far greater.

This also applies to the disruptive discrepancies observed with navigational VBPs: such discrepancies would not be considered as troublesome if it were possible to use reasonably priced motion simulation interfaces that offered sufficiently high technical levels. On the other hand, discrepancies are only very disruptive when the navigation produced by designers is **unnatural and unreal**, that is when users see themselves moving in the VE through their VR headset (movements seen from above, a third-person view). While moving, users can also watch themselves moving in the VE but

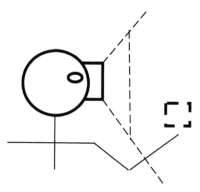

Figure 8.9a Manipulation without seeing the hands in the VE (RE: continuous line and VE: dotted line).

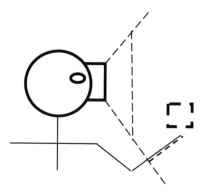

Figure 8.9b Manipulation when seeing the hands in the VE.

can also look elsewhere. This is described as observation with a spatial visual-motor discrepancy, though here with a shift of the point of view in the VE.

For manipulation VBPs, once again in this case a visual-motor discrepancy will not be considered as troublesome if, on a technical level, the user's actual hands are located in exactly the same place as the hands the user is seeing via the VR headset; however, the hands must be represented visually (otherwise, the example detailed in Figure 8.9a applies). The correspondence between the location of the person's actual body (or a part of the body) and the user's visual localisation of his body is called **co-localisation**, and this is an important notion in virtual reality. Co-localisation is the superposition of the visual space over the manipulation area for users of the VR application (see Figure 8.9b). Technically, this co-localisation is currently feasible if designers carefully select the correct hand-location sensors and if the view in the VR headsets is correctly configured and calibrated. However, if you manipulate a virtual object by "tele-operation", that is when seeing an object out of hand's reach in the VE and manipulating it remotely with your hands in the RE, this results in an **unnatural and unreal manipulation** but one allowed by VR techniques. For such tele-operated manipulations, the visual representation of the hands in the VR headset can be achieved at the level of the hands in the RE, that is via co-localisation (see Figure 8.9c); however,

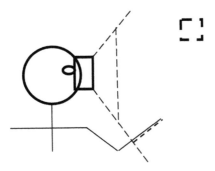

Figure 8.9c Tele-operation with co-localised hands.

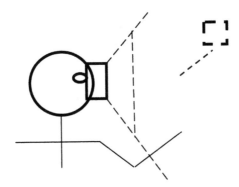

Figure 8.9d Tele-operation with the hands in the VE seen close to the object.

the visual representation of the hands (realistic or symbolic) can also be achieved in the VE at the level of the object being manipulated (see Figure 8.9d). We need to test a case that could make it easier to adapt to this type of tele-operated manipulations.

We can therefore establish a list of disruptive sensorimotor discrepancies according to their cause:

–   Technical difficulties (Dtec);
–   Unnatural and unreal VBPs, imposed by the application's designer (VBP Ur).

The two causes are not mutually restrictive.

–   Observational VBPs:

Temporal visual-motor discrepancy	*(Dtec)**
Visual-temporal discrepancy	*(Dtec)**
Oculomotor discrepancy	*(Dtec)*
Visual-spatial discrepancy	*(Dtec)*
Localisation visual-motor discrepancy	*(Dtec)**
Spatial visual-motor discrepancy	*(VBP Ur)*
Passive visual-motor discrepancy	*(VBP Ur)*

– Navigational VBPs:

Visual-vestibular discrepancy	*(Dtec) or (VBP Ur)*
Temporal visual-vestibular discrepancy	*(Dtec)**
Visual-postural discrepancy	*(Dtec)**

– Manipulation VBPs:

Visual-manual discrepancy	*(Dtech) or (VBP Ur)*

On the list, Dtech* indicates discrepancies linked to the technical shortcomings of the VR heads and which, in view of current developments, are in the process of being eliminated; namely, an adequate reduction in latency time lags, sufficient increase in display frequencies and more accurate tracking of the head and hand. On the other hand, we will need to wait for several years before stereoscopic screens with no accommodation-vergence constraints and VR headsets with a horizontal and vertical field of view identical to those of man are available on the market.

We will explain in the following chapter the methods used to overcome the technical shortcomings in the current VR headsets, as well as the procedures used to manage the discrepancies imposed by designers when they program unreal sensorimotor actions (VBP Ur). Users find it harder to accept them due to the unnatural vision imposed. It should be remembered that virtual reality does not necessarily imply a realistic simulation. Simulations can be unreal and imaginary both in the VE and at the level of unreal sensorimotor actions (VBP Ur). However, in the latter case, users will need to guard against the risks of discomfort and the sensation of feeling unwell.

8.4 CONTROLLING ACTIONS

In terms of cognitive immersion and interaction, users must be able to use cognitive processes easily in relation to sensorimotor interaction and immersion. In particular, users must be able to control their (inter) action in the VE. This is possible only when the latency is low enough to allow users to correctly anticipate their actions according to the sensory stimuli received, as is the case when using a mouse. When the latency is high, users are required to slow down their gestures to control their actions. This changes their behaviour and their natural immersion in the virtual environment will be less extensive. If the latency is too high, this can also cause instabilities in terms of users controlling their actions, which can induce discomfort, especially if it is coupled with other sensorimotor discrepancies. It is currently difficult to determine the combined influence of these two phenomena on the discomfort of users and their sensation of feeling unwell. For better control of (inter) actions, it is preferable that the latency remains constant, to allow users to better anticipate their actions. A learning period may also be required in some cases to allow users to adjust to an excessively high latency lag.

The impact of a high latency lag will differ according to the VBP concerned. A virtual object can be manipulated in two major ways:

– manipulation by seeing the object at the level of his/her hands with a co-localisation between the real and virtual hand observed via the VR headset (and, obviously,

also the co-localisation between the user's actual and the correct viewpoint in the headset);
–   manipulation of a virtual object by "tele-operation" when seeing a virtual object that is out of hand's reach and manipulating it remotely (unreal manipulation allowed in VR) although still with a co-localisation of the hands and head.

For manipulation VBPs with a co-localisation of the hand and head (Figure 8.6b), this co-localisation will be considered as functional if the sensors tracking the head and hand are fairly accurate and the co-localisation is relatively well calibrated. However, this is not sufficient, as the latency lags of the two sensors must be low enough to allow users to control their manual actions. In general, the two *tracking* sensors use the same technology and therefore typically their latency will be of the same magnitude. We can assume (although this is yet to be confirmed) that a lengthy lag with a hand tracker will be particularly troublesome for controlling manual actions (due to the hand-eye coordination) and that a lengthy lag with a head tracker will cause discomfort and make users feel unwell due to the temporal visual-motor discrepancy. For manipulation VBPs via tele-operation (Figures 8.6c and 8.6d), in principle, controlling the actions will be harder when the latency lag is too long, regardless of whether or not the observer can see his/her co-localised hands (or via a simple visual cue). In terms of tele-operation, we know that the positioning and rotation of objects are harder movements to perform cognitively when "you handle them in your hands". We have also checked this, along with others, and a small percentage of people may struggle to position a manipulated object using a 6 dof sensor in the real environment, while looking at the object in the VE remotely.

Observational VBPs are easier to analyse than the previous case: all that is required is to be sufficiently accurate and, especially, that the sensor tracking the head should have a short latency lag to be able to easily control the point of view.

In terms of navigational VBPs, several types of latency come into play: the sensor tracking the motion control interface (in general, manually controlled), the sensor tracking the head and, in some cases, the one used by the motion simulation platform. These latencies must be low and constant to allow users to correctly anticipate and therefore properly control their navigation in the VE.

8.5   PSYCHOLOGICAL PROBLEMS INDUCED BY VIRTUAL ENVIRONMENTS

VR headsets will be used for professional and leisure activities, especially for video games. In general, professional use will be well targeted and limited, and should not pose any specific psychological problems. However, virtual environments will probably – although this is not yet certain – be used for lengthy periods during gaming and recreational activities. Since users will be immersed in panoramic photographs or 360-degree videos when visiting a place, the duration of use should be relatively short and limited. Therefore, the risk of addiction will be minor. However, when a VR headset is used for video games, players can potentially spend more than one hour a day gaming. Psychological consequences can occur from excessively long periods immersed in a virtual world. The fundamental question worth asking is whether the risk of becoming addicted to video games is higher if the player uses a VR headset instead of an ordinary screen? This question addressed to psychologists and psychiatrists has not yet been debated, as this type of usage is still not widespread.

The issue is to determine whether when gamers look at a VE via a VR headset they are more immersed than if they were sat in front of a computer screen or games console? In principle, gamers would be in the same situation, favouring a lack of socialisation, confusion between the real world and the virtual world or leading to addiction. Wearing a VR headset cuts off the gamer's view of his or her real environment. We will analyse below the consequences in terms of the wearer's physical safety. However, gamers sat in front of a screen, usually in the dark, are equally prone to cognitive isolation. The degree of immersion induced by virtual reality headsets is not the only factor responsible for addiction: the design of scenarios in the video games also triggers a certain type of addiction as gamers will constantly strive to have more experience to become the best.

Video games, as well as any other virtual-reality experience, can have an impact on the imagination of young children, particularly a psychological impact. The issue is to determine whether an intensive use of VR headsets could have a specific impact on the psychology of children that differs from the one resulting from visual immersion via a screen (when such a use is permitted in children, which is contrary to the recommendations of manufacturers of VR headsets). This chapter, however, will not debate this important issue.

8.6   OPTICAL AND ERGONOMIC CONSTRAINTS

Ophthalmologic problems may occur as a result of users seeing a virtual scenario via an optical device that is not specially tailored to their own eyesight. Unlike the situation with correctives glasses, ophthalmologists and opticians have no involvement in the use of VR headsets. Consequently, headsets are not adjusted to the eyesight of each wearer, despite the fact that a percentage of the population has problems with their sight. The optical constraints video gamers may face by wearing VR headsets of varying quality is an issue worth considering. These types of headsets offer few settings allowing them, at the very least, to be adjusted to the morphology of users. On top of this, the few optical adjustments and calibrations they do offer will be checked only very infrequently. With such checks carried out at the discretion of users and without any involvement of the professionals.

The ergonomic problems are well known: the weight of the VR headset can induce an increase in physical symptoms when the subject wears a heavy helmet for too long. However, given that the equipment is becoming increasingly sophisticated, this problem can potentially be avoided by wearing a lighter helmet. VR headsets and their fastenings must be adjustable to any head shape using fairly effective technical solutions. It is quite amusing to see on demonstration videos, a person removing his hands from the VR headset to manipulate a virtual object and then replacing nimbly his hands on the VR headset, which looks like that it needs help to stay on his head. Another constraint is the heat users may feel on their face. Poor ventilation in the VR headset can also cause fogging on the lenses on some models.

The cables connecting the computer to the VR headset can also hinder certain of the person's movements. Cables are set to remain for some time on high-performance VR headsets, while those used with smartphones do not have any. One solution for a person who is standing and is required to swivel around could be to wear a computer on a belt – a solution particularly favoured by VR headsets used for Augmented Reality. Designers need to take into account the cables and the potential hindrance to the person's movements. Cables could descend from the room's ceiling before reaching the VR headset to be less restrictive.

There are several arrangements for individuals using VR applications, the main ones are as follows:

– the person is sitting on a fixed seat;
– the person is sitting on a seat swivelling 360°;
– the person is sitting on a motion simulation platform;
– the person is standing and is motionless;
– the person is standing and moving around the room;
– the person is walking on a treadmill;
– the person is stretched out on a swing seat.

The ergonomic constraints will differ according to the arrangement selected.

The communication with the configuration of the VR application, including that with the VR headset, must also be analysed. Communication is generally not highly ergonomic as users do not have access to an ordinary keyboard and a screen. They often have to perform commands with buttons located on the headset or on levers, without seeing them.

We will explore below some of the solutions proposed to reduce or even eliminate sensorimotor discrepancies. Such problems could be resolved on a technical level or through the equipment used that is by using Behavioural Software Aids (BSAs) or by changing the Virtual Behavioural Primitives (VBPs). We established our classification by decoupling each sensorimotor discrepancy for greater clarity; however, these discrepancies may mutually interact. In reality, a person may have the sensation of an overall discrepancy but is not generally aware of the exact place where each sensorimotor discrepancy is located. Indeed, many people are totally unaware of the presence and effectiveness of their vestibular systems.

Chapter 9

Recommendations and solutions

This chapter outlines the ways used to minimise the technical shortcomings in the current VR headsets and the methods used for handling the discrepancies imposed by unreal Virtual Behavioural Primitives (VBP Ur). Users find it harder to accept them due to the unnatural vision imposed.

Up to now, very few experimental studies on the use of VR headsets by the general public have been carried out. However, some research into sensorimotor discrepancies observed with professional VR applications has been conducted and, more historically, research has been carried out on transport simulators. A VR headset is not widely used with these VR applications and transport simulators. Users face identical discrepancies, whether they are in front of a fixed screen or are wearing a VR headset. As long as the extrapolation is justified, experimental studies can be the source of solutions and recommendations, and they are analysed in the context of the "Perception, Decision, Action" loop. We analyse each sensorimotor discrepancy independently to simplify the analysis. However, the senses may be coupled together when analysing the disruptions suffered by users, even if they are not taken into account. Similarly, two VBPs are often executed at the same time as visual observation is permanent. Users are able to perform two motor actions simultaneously such as, for instance, walking on a treadmill and turning their head while also observing the VE through the VR headset. However, we will study the VBPs separately to simplify the analysis. We present solutions and recommendations on the basis of the list of disruptive sensorimotor discrepancies and, for each one, will ask **four questions**:

– How can you minimise the impact of sensorimotor discrepancies in terms of the discomfort of users or the sensation of feeling unwell?
– It is possible to eliminate the sensorimotor discrepancies by modifying the way VBPs work?
– Can we eliminate sensorimotor discrepancies by changing the way the behavioural interface works or by adding another behavioural interface?
– How can users adapt to discrepancies so they no longer experience discomfort or have the sensation of feeling unwell?

The first three questions will be asked for each case of discrepancy. We will then tackle the issue of adaptation comprehensively. The list of 11 disruptive sensorimotor discrepancies and their category (see chapter 8), as well as their cause due to either technical difficulties (Dtec) or an unreal Virtual Behavioural Primitive, imposed by the application's designer (VBP Ur). We present 32 solutions, ranging from the most classic to others very rarely used or even not yet totally validated. The objective of the following paragraphs is to establish the most comprehensive list possible. We will then propose an analysis grid that will allow us to determine the solutions, as we need to perfectly understand and control some of these, while others will be for specific usages.

9.1 OBSERVATIONAL VBPS

9.1.1 Temporal visual-motor discrepancy

This is a problem specific to all VR headsets: the latency between the observer's head movement and the correct display of the point of view corresponding to the movement of the VR headset.

Reducing the discrepancy (S1)

Several approaches are possible to lessen the impact of this discrepancy:

–   we can reduce the latency lag *on a technical level* by improving the execution times of programmes, by changing the computer, the graphics cards, the location sensor or the VR headset itself;
–   if the latency lag cannot be shortened enough, we can use algorithms to predict head movements (Kalman filter, etc.). However, sudden head movements mean that the prediction will obviously fail. Some suppliers of VR headsets, such as Oculus, offer predictive algorithms whose use we recommend.

Notes: details of solution S1 is given after solutions S2 and S3.

Removing the discrepancy by changing the way VBPs operate (S2)

This is a rare and unusual use of VR headsets and one that may be problematic for observers who feel extremely uncomfortable inside a VR headset; this solution concerns highly specific VR applications in which images are only displayed when the head is immobile, thereby placing very few demands on the observer. Any rotation of the user's head will make the images turn entirely black or appear in a homogeneous colour that corresponds to the average shade of the virtual scene. It is worth noting, however, that with this particular display of the sequence of images, a latency lag is observed when changing from the stationary phase to the motion phase, and vice versa, although these transitions are less disruptive and can be achieved gradually by fading out the images. A sensor tracks the stillness of the head and the measurements it takes are filtered (which involves a delay) to remove any parasite movements as long as the user wishes to retain the same viewpoint.

Example of specific use: looking at a place or a building using a certain number of different viewpoints without needing a dynamic immersion of the gaze and without moving in the VE, to not overtax the observer. Movements in the VE can be achieved indirectly by, for instance, using a given position on a map, etc. though, obviously, the visual immersive experience is much weaker. This solution can also be used merely on a temporary basis to allow observers to rest their visual system.

Figure 9.1 The immersed-gaze functionality can be achieved using three U-shaped flat screens surrounding the observer's head.

Removing the discrepancy by adding/modifying behavioural interfaces (S3)

To eliminate the latency lag, on the condition that the user's head does not change, or virtually does not change, its translational movement, a VR headset will no longer be used but the VE images are displayed on one or more fixed screens in the real environment. To retain the immersed-gaze functionality, screens should surround the user's head, which can be achieved using a parabolic screen or with three U-shaped flat screens surrounding the observer's head (Figure 9.1). The images are calculated independently from the rotation of the head. Images will be correct if the head remains motionless or almost motionless in terms of translational movements, which is often the case when the person is seated in front of the display device. If this is the case, we no longer need to measure the rotation of the head and the discrepancy is eliminated. In addition, the quality of the images may be better, as an extremely high number of pixels can be displayed. These solutions, which are comparable to that of a good quality VR headset in terms of cost, could be interesting for some applications in which the user's head remains in a relatively fixed position, while still retaining the freedom to turn.

Details of solution S1

In virtual reality, the latency time (or simply latency) is the time lag between the user's action and the response perceived, as a result of his action. This latency results from adding several response times together: the motor interface (joystick, force feedback interface, etc.), the computer processing of the control signal received, the time it takes the computer to create the response required (image rendering, etc.) and the transfer of the response to the sensory interface (in our basic case, the time required to display the correct image in the VR headset) (Figure 8.2).

The latency is therefore dependent on at least two different types of equipment that may or may not be supplied by the same manufacturer. A head *tracking* sensor is generally included in the VR headset. The latency data supplied by manufacturers can often be minimal and only achievable in practice using a computer with the same power and when running an extremely short programme. It is worth restating that very high frequencies are required for head *tracking* and displaying images in the VR headset. It is a necessary but not sufficient condition. For example, for infrared cameras that detect the location of the head with a frequency of 50 Hz, a measurement can only be taken every 20 ms. The latency will automatically be higher than 20 ms. As this is the order of magnitude that needs to be attained, movements must be tracked at frequencies higher than 50 Hz.

Given that the vestibule-ocular reflex occurs at approximately 20 ms (Berthoz, 2002), many VR specialists believe that a 20-ms latency lag is the target to be reached. And lower than 10 ms would be highly desirable. A latency of over 50 ms is too high and the lag between the head's movement and the displaying the correct viewpoint is far too long. The virtual environment will not appear stable and this may cause problems when controlling the viewpoint. Although as yet no clear rules have been determined regarding the latency lag to be attained. Some research advocates that the latency should be constant, and therefore predictable and controllable after a certain period of adjustment (Draper, 2001).

How can we reduce latency? This is a critical problem for manufacturers of VR headsets and the associated (peripheral) sensors. For developers of applications, the question should rather be: How can we not increase latency too much? The programme must be well designed and developed to ensure that the time between reading the data originating from the head-tracking system (and possibly the hands) and the rendering of images is not only shorter but also constant. The frequency of the image rendering cycle must be similar to the frequency of the screen's VR headset that is at least 50 Hz. To avoid increasing the latency, the complexity of the VE should therefore be limited, and actions made parallel.

It is advisable to check the latency measurements, both for the time constraints linked to the equipment as well as those relating to the programme. Oculus supplies a "Latency Tester" device that measures the latency with a given scenario. The device is placed on one of the lenses of the VR headset to detect any change in images. The method they suggest is as follows: you need to press a button on the tester to start the test; a series of messages are then sent and interpreted by the rendering system (such as, for instance, an order to change the colour of the screen). The time elapsed between sending the message and the screen actually changing colour is measured for each message, and then averaged out to obtain a latency measurement. As the length of messages and speed of communication are known, the latency can then be calculated. This device does not measure the response time between the actual movement of the head and its detection by the VR headset.

We are proposing an independent measurement system for manufacturers that can be used on all VR headsets, and which we have already used for VR applications on a screen or Cave. It requires a very high frequency camera (approximately 1000 Hz) to obtain an accuracy of 1 millisecond to measure the overall latency. For on-screen VR applications, the camera will simultaneously take pictures of the displayed images and the hand (or head or feet) of the person, depending on the latency being measured. This is easy to do with a screen; however, it needs to be used more carefully with a VR headset, as it cannot be worn. The camera is unable to film images in the headset but can detect at least one colour change coming out of one of the two lenses. By including in the programme the instruction to change the colour of the scene in the VE when the VR headset is rotated, the camera will film the start of the movement in the VR headset and then the moment the colour changes. Latency is measured by the number of images occurring between these two times. We may also want to measurement the latency of hand commands, in relation to hand movements. If this is the case, the camera will film both the hand and output from one of the lens. This external method is a little complicated to implement but offers a fairer and independent evaluation of both the hardware and software.

9.1.2 Temporal-visual discrepancy

A disruptive discrepancy occurs if the frequency of images displayed is too low compared to the needs of the visual system to perceive flicker-free images with continuously moving objects in motion.

Reducing the discrepancy (S4)

On a technical level, manufacturers need to use screens with the highest frequency possible, which has been gradually taking place in recent years and this trend is set to continue. In the meantime, it is possible to process the images using spatial and temporal filtering or by not creating content that causes flickering: for example, textures with high spatial frequencies for moving objects, very fine objects (grid, meshes, etc.), repetitive-pattern textures, rapid light alternations, etc. If there is text to read on objects, the font should be enlarged to make it legible. Such conditions can be checked by testing. These recommendations are currently important as the definition of the screens on VR headsets is relatively low. It is already possible with many of the current VR headsets to see the pixels, which is fairly unpleasant. If, in addition, the pixels are flickering, the quality of the image will be even poorer.

It is possible to filter out parasite movements (tremors) of the observer's head and the noise from the head-location sensor; we can partially remove the jitteriness of images by freezing their viewpoint, as long as the observer does not change the direction of his viewpoint, by using a spatial-temporal and filtering mechanism: if, over a short period of time (around one second), the axis of direction and the centre of the head do not vary by a small angle and a small translational shift, and the viewpoint of images is fixed. It will not be fixed as soon as the head exits the neutral zone. The negative aspect of this filtering is that it partially increases the latency and that it is not effective for moving objects in the VE scene. Furthermore, the distance of textured objects relative to the camera is a parameter that can influence the flickering of pixels. This can be taken into account to improve image quality.

Removing the discrepancy by changing the way VBPs operate (S5)

In common with solution S2, images are not displayed continuously but only when the head is motionless. The viewpoint is frozen as long as the head does not move. When there are moving objects in the scene observed, partial flickering will occur. The discrepancy is therefore only partially eliminated and, as indicated above, this solution is only rarely used.

Removing the discrepancy by adding/modifying behavioural interfaces
Inappropriate request.

9.1.3 Oculomotor discrepancy

In stereoscopic vision, the "accommodation-vergence" discrepancy becomes a hindrance when the retinal disparities imposed on observers exceed a certain limit, which is dependent on the visual capabilities of the observer, the exposure time and other criteria.

Reducing the discrepancy (S6)

By reducing retinal disparities, and sometimes also by processing the images, we can help observers to merge the stereoscopic images thereby exerting less strain on their eyes. In addition, if the optical axes of the eyes are tracked in real time in the VR headset, the processing of images will be more effective as it will be possible to determine precisely at all times the exact place where the observer is looking in the VE. At present, very few VR headsets offer eye-*tracking* systems.

   Notes: details of solution S6 is given after solutions S7 and S8.

Removing the discrepancy by changing the way VBPs operate (S7)

This simply requires displays offering a monoscopic vision. This option should be provided systematically for the small percentage of the population who are unable to see in stereoscopic vision and for people who really struggle to merge stereoscopic images and therefore place considerable strain on their eyes. This discrepancy is less important than with conventional stereoscopic screens, as monocular cues, particularly the motion parallax (a new functionality offered by VR headsets), allow users to perceive much better the three dimensions as compared with stereoscopic vision alone. Obviously, it is best to combine these two depth perception indices when observers have no problems with stereoscopic vision. Changing to a monoscopic vision is therefore not a very strong constraint. Users will only experience some problems when they need to manipulate delicately objects that are located near their head. In this case, carefully chosen Behavioural Software Aids can help users to achieve the manipulation VBP. Monoscopic vision can also be used on a temporary basis in some situations that overtax the observer's visual system (see S6 solution).

Removing the discrepancy by adding/modifying behavioural interfaces (S8)

The use of Light Field technology screens or VRD "Virtual Retinal Display" screens can eliminate this discrepancy, since, in principle, with these types of screen, the accommodation and vergence occur at the same correct distance for each point of the VE observed. Detailed explanations of these types of screens can be consulted in the book "Displays, Fundamentals & Applications", by Rolf R. Hainich and Oliver Bimber[1]. The first VR headsets using this kind of technology are expected in 2016, but will offer relatively limited fields of view.

---

[1]ISBN 978-1-56881-439-1, publisher CRC Press.

Details of solution S6

There are several ways to reduce the "accommodation-vergence" discrepancy; the main method is to limit retinal disparities below a certain threshold. One method consists of setting the threshold at a precise and definitive value for the entire VR application (1.5°–1.2° or 1°, for example) (see (Shibata, *et al.*, 2001)). However, we know that the ability to merge images with ease depends on the type of images observed, as well as the duration of the observation and the visibility capabilities of each individual observer. We can also use more refined methods such as:

–   controlling the observation time of retinal disparities and their intensities;
–   processing the pair of stereoscopic images;
–   processing the pair of stereoscopic images by determining where the observer is looking, which is feasible when eye-tracking VR headset are used;
–   adjusting to the capabilities of each observer, given that, as we have already stated, for some people, the threshold is zero, meaning that a monoscopic-vision option must be provided for such users. A monoscopic vision does not prevent space from being perceived in three dimensions; however, the perception of depth will be less effective. All the monocular cues of depth perception will still remain.

The principal method is to limit the sensorimotor discrepancy in order to not display images with a retinal disparity greater than a threshold of around 1° to 1.5°. Disparities will be created by positive or negative horizontal parallax angles (crossed or non-crossed) on the actual screen or virtual screen, in the case of a VR headset. A horizontal parallax is the angle between two points corresponding to the left and right images displayed on a screen, viewed from the eyes of the observer. Consequently, for all VR headsets, we need to determine the location of the virtual screen relative to the position of the eyes, i.e. the accommodation plane or the focal plane (it is around 1 m or 2 m depending on the VR headset). For the DK2 Rift headset from Oculus, the distance is approximately 1.3 m. This company recommends that objects are positioned at a distance ranging from 0.75 m to 3.5 m, in order to limit retinal disparity (Figure 9.2).

The horizontal retinal disparity is approximately equal to the horizontal parallax angle, which means that in practice we can use formulas to set limits that should not

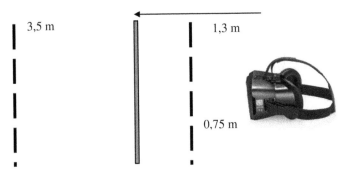

3,5 m

1,3 m

0,75 m

Figure 9.2 The objects are positioned at a distance ranging from 0.75 m to 3.5 m, in order to limit retinal disparity.

be exceeded. As the virtual screen is located at a distance of 1.3 m, an object located at a distance of 0.75 m will imply a retinal disparity δ:

For an interpupillary distance (IPD) = 0.065 m, with P as the horizontal parallax for the homologous points of an object at 0.75 m, then we have P = $(1.3–0.75) \times 0.065/0.75$ and tg $(\delta/2) = P /(1.3 \times 2)$, which gives 1.05°. Using the same calculation, we obtain for an object 3.5 m away, a retinal disparity $\delta' = 0.9°$.

This company therefore recommends limiting retinal disparities to around 1°, which is a satisfactory limit for a wider audience. In comparison, for 3D cinema, the retinal disparity for spectators seated in the front rows, and who are the most exposed to the discrepancy, is around half, as the duration of exposure also has a considerable impact on the observer.

Designers using stereoscopic vision in their VR applications (whether or not they require a VR headset) may pay close attention to the principle of limiting the viewing area by restricting the front and back of the screen. However, this limitation is not absolute, as it is always possible to temporarily position objects either further forward or behind the zero parallax area. These will then be displayed with very large horizontal parallax and will therefore be observed with extremely high retinal disparities. This is what happens in 3D cinemas when, from time to time and for short periods, objects seem to spring out of the screen to the delight of spectators. For 3D cinemas, stereographers determine a sort of "stereoscopic budget" for the entire film to avoid overtaxing the visual system of the spectators. The lengths of each sequence with each level of disparity are counted and should not exceed a certain limit. It is worth considering a similar procedure for any VR application that lasts for longer than 15 minutes.

Steps must be taken to ensure that observers do not observe very close objects with a disparity greater than 1.5° for too long. This may be the case for example, when a user looks through the viewfinder of a weapon located to close to his head. In such situations it is better to transfer to a monoscopic vision. As the observer is concentrating on the viewfinder, they will not really perceive the relative flattening out of the VE as a result of the removal of the stereoscopic vision. Although the transition should be a gradual one. Another example of this type of situation is when information is given to users and displayed in the foreground of the scene. If the display plane is located very close to the observers' eyes and is viewed in relief, they will become tired. Moving the plane of the display screen back will make it easier to see objects in the virtual screen that are closer to the eyes; however, they should not be hidden as this would result in a conflict between the monocular occultation index and the binocular depth-perception index. There is no ideal solution, as is the case for the subtitles on stereoscopic films for which zero disparity is used to prevent eye strain for spectators and also to avoid concealing any objects jutting out from the screen. It may be appropriate to display information from the VR application on the plane of an object, if the scene makes it possible.

Monoscopic vision can therefore be used temporarily, for certain situations that overtax the observer's visual system, whenever we know that the user is going to place an object close to his eyes for a certain period of time or when the application's scenario imposes such an action. Oculus recommends this approach when objects are located at a distance of less than 0.5 m, which is consistent with the characteristics of their VR headset. Consequently, there is always an option to switch to a monoscopic vision or at least lowering the stereoscopic vision artificially by reducing the retinal disparity.

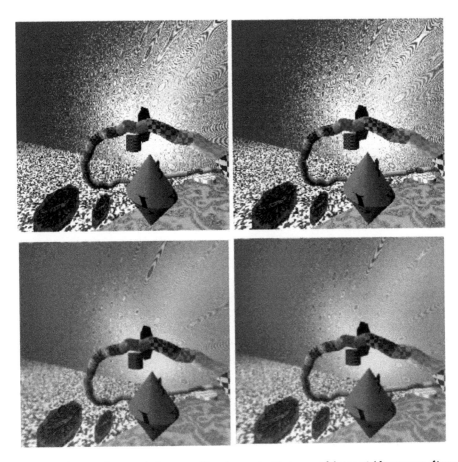

Figure 9.3 Reduction of eye strain by controlling the content in terms of the spatial frequency of images: stereoscopic images without (top) and with "Perrin comfort function" (below): deletion of high spatial frequencies where the retinal disparities are high (The accommodation plane is positioned on the object which is in front of images).

Another consideration to be taken into account is whether objects are in motion. The observer's visual system will be required to merge pairs of stereoscopic images more rapidly. The retinal disparity threshold must be lower when objects are in motion.

Similarly, more restrictive conditions apply when the spatial frequencies of the images are high. We can reduce eye strain by controlling the horizontal parallax between the left and right images and by controlling the content in terms of the spatial frequency of images (Figure 9.3). Reducing eye strain is dependent on the spatial frequencies and retinal disparities, and the use of a visual comfort function in the image. This function means that it is possible to determine if a point in the image is tiring the observer and if the image point requires processing. The higher the spatial frequencies and retinal disparities in one point of the image, the greater the eye strain will be. The comfort function was determined experimentally in the work carried out

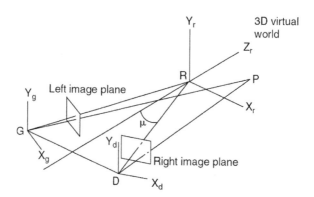

Figure 9.4 The two optic axes of both virtual cameras converge on the main viewing area.

by several researchers (Perrin, 1998). Without using this comfort function, designers of stereoscopic images can at least blur the areas where the retinal disparity is high.

Tracking the observer's eye movements will help improve the processing of stereoscopic images and enhance visual comfort for users: this means that images will be sharp in the area being viewed while those in the peripheral vision will be blurry, thereby eliminating the high spatial frequencies that are troublesome when the retinal disparities are high. Some VR headsets will soon offer eye-movement tracking as an optional feature.

Overall, it should be noted that the proposed stereoscopic vision is an artificial vision with limitations as it imposes a fixed distance on the accommodation plane, and on the plane of the real or virtual screen. This is a major difference compared with natural vision in which the distance of accommodation varies constantly. An interesting feature would be the option to change the distance of the accommodation plane, the focal plane of the VR headset, depending on the dimension of the virtual scene in a given application. However, the current crop of VR headsets do not provide this option, as optically it would be too complicated to implement. The future of VR headsets, as previously stated, lies in being able to technically remove these accommodation-vergence discrepancies through the use of Light Field screens or Virtual Retinal Display (VRD) technology.

Stereoscopic vision has been the subject of research for a long time (the first recommendation dates back to over a century and a half ago), with a number of well-established reports detailing a range of different fields such as three-dimensional photography, 3D cinema and television (and non-3D), virtual reality and stereoscopic natural vision, and which all place limitations on horizontal retinal disparities.

What criteria should we rely on to position the two points of view (virtual cameras) generating the stereoscopic image pairs? In principle, it could be postulated that it would be better to converge the two optical axes into the main area to be viewed, to reflect natural human vision more closely. However, in this case, the projections on both planes provide images with horizontal parallax angles but also with vertical parallax ones (the pair of images corresponds, to the nearest multiplication factor, to two images displayed on the screen (Figure 9.4).

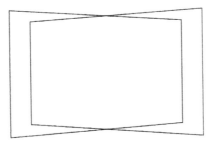

Figure 9.5 Vertical parallax created with the two optical axes of two virtual cameras that converge (the vertical parallax is exaggerated in the figure).

This case should to be avoided, since it imposes a vertical parallax that makes it difficult to merge the images. For example, the left and right projections on the front side of a cube centred on axis Zr give the images shown on Figure 9.5 with vertical parallax angles at point P that will become larger as point P moves away from axis Zr.

Merging is difficult with a vertical parallax greater than 20 degrees of arc. The optical axes of the two viewpoints need to be parallel to avoid having a vertical parallax that causes problems for the display. Algorithms synthesising the creation of stereoscopic images almost always take account the parallelism[2] of the two axes of the two viewpoints (of the virtual cameras).

It is preferable that the spacing between the optimal axes of the cameras is equal to the interpupillary distance of the observer. Although this is possible and preferred for 'normal' scale scenes, the following cannot be considered in extreme cases:

- For stereoscopic aerial views, the spacing is several hundred metres to give an effect of depth, which is not natural but useful. In this condition, objects such as buildings are perceived as being smaller than they are naturally. This is the "model" effect that is well known to stereographers. This use can be considered when piloting drones via a VR headset to offer a better artificial perception of distances; however, it can also cause greater eye strain.
- To view minute objects (e.g. insects), the spacing needs to be less than the interpupillary distance.

In general, with the use of VR headsets, scenes are viewed at scale of 1 and therefore perceptions of dimensions with monocular cues and binocular cues are matched, that is to say the vision is orthostereoscopic. Stereoscopic vision, both natural and artificial in a VR headset, is particularly useful for objects nearby with which you want to interact, which is often the case for VR gaming applications. Hyperstereoscopic vision and conversely hypostereoscopic vision can also be used in various areas, especially

---

[2]When tracking an observer's eyes in real time, the axes of the two cameras can be converged at the observed point, as the vertical parallax angles will then only be created in the peripheral vision, as is the case with natural vision. This method is still in the research stage.

for 360° videos. However, the disparities imposed must be controlled accurately and it should be remembered that the result provides a distorted perception of reality.

An Augmented Reality (AR) headset using a pair of stereoscopic cameras filming the real scene is a specific example of this. In an AR headset, for technical reasons the cameras are only spaced 40 mm apart and not 65 mm. To combine synthesised images with real images, synthesis images need to be created with the two virtual cameras spaced 40 mm apart and not 65 mm.

Under these conditions, eyes that are 65 mm apart on average, will adjust; however, the environment will be perceived as being larger than in reality with a theoretical ratio of 1.6 (65/40) on average. This is only theoretical, as the presence of objects with a known size (such as the hands of the user among others), allows the brain to compensate using monocular cues.

For all types of VR headsets, it should also be noted that they only offer a stereo-scopic vision in a central area of the overall horizontal field of view. In some VR headsets, this central area of binocular vision is relatively low and any sensorimotor discrepancy will obviously have a lower impact, which also applies to the perception of depth and therefore the visual immersion. However, for other VR headsets, the total field of view is entirely in stereoscopic vision, but in these types of headsets the total field of view will not exceed 100° to 110°.

To complete this set of rules to implement for stereoscopic vision in a VR headset, it is also worth talking about the adjustment ability of each user, which can vary considerably. For some people, the stereoscopic vision must be removed totally as they are unable to get used to it. 3D cinema has correctly understood this constraint, as the same film will be projected in both a 3D and a normal mode (monoscopic – which should not be referred to as 2D) – despite the fact that some proponents advocated a few years ago that all films would now be produced in stereoscopic vision. The advantage with VR headsets is that each observer can individually select their preferred option, when they are given a choice by the designer: namely, normal stereoscopic vision, moderate stereoscopic vision or monoscopic vision.

Given that the Interpupillary Distance (IPD) of the observer varies considerably, as previously indicated, it is essential that all VR headsets allows the spacing between the two optical axes to be adapted to the IPD of the observer. And, obviously, if we want the observer to perceive accurately the exact shapes and distances, the IPD value of each observer must be integrated into the stereoscopic-image design software. Some manufacturers of VR headsets are proposing a method to determine the correct settings depending on the Interpupillary Distance (IPD) of the person, and this seems essential for any application lasting longer than five minutes.

To conclude the section on the adjustment of observers to the "vergence-accommodation" discrepancy, it should be remembered that action can be taken on several parameters: the duration and intensity of retinal disparities, the processing of the spatial frequency of images to "soften" them and, in some cases, to offer mono-scopic vision. In the VR application it may be highly appropriate to include a simple test to evaluate the stereoscopic vision capabilities of the wearer of the VR headset, and taking into account the results when setting the stereoscopic vision.

9.1.4 Spatial-visual discrepancy

A disruptive discrepancy occurs if the fields of view of the VR headset fields differ from the fields of view of the camera filming the VE (Figure 9.6) (Draper, 2001). It is

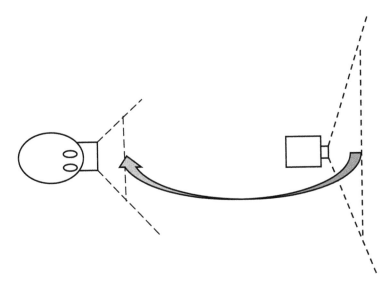

Figure 9.6 The fields of view of the VR headset fields differ from the fields of view of the camera filming the VE.

strongly recommended that the two fields of view should remain the same. However, it should also be noted that this becomes a constraint when using a VR headset instead of an ordinary screen. On an ordinary monoscopic screen, the actual scene can be filmed with a field of view that differs from the observation field of view, depending on the size of the screen and viewing distance. The brain cognitively perceives a virtual world with coherent dimensions. If the display is stereoscopic, the monocular and binocular cues must be coherent (orthostereoscopic vision) to perceive a coherently dimensioned world. Alternatively, in hyperstereoscopic or hypostereoscopic vision, dimensions are transformed due to the "model" effect, as explained above. By amplifying the field of view artificially, the world observed becomes more incoherent. The vestibulo-ocular reflex is inactivated and users will have to adjust to this situation, which is extremely demanding. Users may have great difficulty when manipulating virtual objects as the co-localization of their hands will be inactivated. We therefore need to carefully consider the issues involved before artificially increasing the too-narrow field of view of a VR headset. The solution may be counter-productive.

Reducing the discrepancies

Inappropriate request.

Removing the discrepancy by changing the way VBPs operate (S9)

The fields of view are equalised, even if the VR headset's field of view is very poor. This discrepancy no longer exists but the view is not very immersive, especially for VR headsets with an ultra-low field of view where users appear to see the scene as if they were looking through a tube. Limiting the field of view is challenging for a visual functionality that uses the peripheral vision; when an object or any other entity is moving in the peripheral vision, the observer is alerted by this movement, even if it cannot be seen clearly (greater distribution of rods on the periphery of the retina). The observer has to turn his head to better see the moving object. When the field of view of the VR headset does not allow this alert functionality, a subterfuge can be used; using the headphones, it is possible to create a surround (3D) sound on the side of the observer's head, at the precise location of the virtual object where the observer needs to look. Another solution is to draw the eye visually sideways by, for example, placing an object that draws the eye to it (flashing, etc.), and half-hidden at the limit of the field of view. The observer will tend to partly turn his gaze to the object and so forth, when necessary.

Removing the discrepancy by adding/modifying behavioural interfaces (S10)

A "full-face VR headset" such as StarVR can be used. A VR headset is called "full-face" when its fields of monoscopic and stereoscopic view, both horizontal and vertical, are just as large as those of the visual system of human beings.

9.1.5   Localisation visual-motor discrepancy

Head movements in the real environment control the visual rotation in the VR, which may differ as the sensor may be inaccurate for rotational movements or because the sensor does not measure translations.

Reducing the discrepancies

Inappropriate request.

Removing the discrepancy by changing the way VBPs operate

Inappropriate request.

Removing the discrepancy by adding/modifying behavioural interfaces (S11)

We can use a head localisation sensor that is sufficiently accurate and measures the 6-DOF: 3 rotations and 3 translations.

9.2   "UNREAL OBSERVATIONAL" VBPS

9.2.1   Spatial visual-motor discrepancy

The designer of the VR application may wish to program an unnatural visual observation: rotational or translational amplification, or a viewpoint unrelated to the head (third-person vision).

Reducing the discrepancy (S12)

We can limit the increased amplification, for rotations as well as for translations, to not move too far away from natural vision. For rotations, if the amplification is designed to allow the observer to perceive a larger field of view without forcing him to turn his head too much (for example, when cables hinder such movements), another solution is to provide control via hysteresis or a non-linear control.

Hysteresis control

When the observer turns his head a little, the relationship between the head's angle and the angle of view is linear (normal operation of the VR headset). However, when the observer turns his head at a certain angle $\alpha(45°–60°)$, the angle of the viewpoint rotates with the amplification. When the observer then turns in the opposite direction, the normal linear relationship is re-established, as the head's angle of rotation has not reached angle α. This offers a normal vision over a certain narrow angular area and, when the user wants to look at another area in a direction that is further away, the area will change using this particular control without the user having to move his head and body too much. Therefore, the observer changes the visual orientation in the VE without changing the orientation of his body in the real environment (which is useful if the user is seated and wants to look behind himself – see also the following control). It will obviously take some time to adapt to this unnatural visual control.

Non-linear control

This involves the same principle but as soon as angle α is exceeded, the rotation of the head controls the rotation speed of the viewpoint rotation angle. The visual orientation in the VE in relation to the orientation of the body in the VE will change must faster.

Removing the discrepancy by changing the way VBPs operate (S13)

We can minimise the disruption to the third-person view by displaying this view on a virtual screen, which is included in the field of view on the VR headset, and is fixed in relation to the headset, and therefore, the head. The viewpoint is changed manually and not by a head movement, as if the observer was merely acting in front of a flat screen, apart from the fact that the virtual screen is connected to the observer's head. This type of third-person view on a virtual screen can be used if it is included in the normal viewing window on the VR headset, to allow the observer to better visualise the scene; however, it will require a greater visual effort, as it is unnatural. For example, the third-person view of a place seen from above, controlled manually, while simultaneously also seeing the place in direct vision, controlled using the head. It offers a vision with two simultaneous viewpoints, which is very artificial and unreal but which is permitted by a VR headset, especially if it offers a large field of view and high resolution. This vision is also disruptive but differs from the vision controlled by the head in the third-person view as, in this latter case, the risk of not understanding the command and causing disorientation is greater.

Removing the discrepancy by adding/modifying behavioural interfaces

Inappropriate request.

9.2.2 Passive visual-motor discrepancy

Visually speaking, the VE is in motion (vection) **without** the user, who is motionless in the RE, moving his own head. This discrepancy is extremely disruptive as it does not correspond to natural behaviour and should only be used rarely and with care.

Reducing the discrepancy (S14)

Motion in virtual environments, when not controlled by the user, must be minor in terms of speed and acceleration, and this also applies to translation and rotation movements. Recommendations on speeds, accelerations and types of trajectories will be explained in greater detail below in the paragraph on visual-vestibular discrepancies when the movement is controlled by the user. As we have already indicated, if the person is passive and not active, like a passenger in a vehicle, limitations on speeds, accelerations, must be greatly reduced and this type of discrepancy is strictly reserved to users who have adapted to such demands. For uncontrolled public use, this discrepancy should be avoided.

Another solution is to change the viewpoint intermittently, in stages and by removing totally or partially the vection. Changing from one position to another viewpoint can also be achieving by fading out.

Removing the discrepancy by changing the way VBPs operate

Inappropriate request.

Removing the discrepancy by adding/modifying behavioural interfaces

Inappropriate request.

9.3   NAVIGATION VBPS

To simplify the explanations, in the following paragraphs we will not differentiate, as was the case for observational VBPs, discrepancies due to the technical failings of the VR headsets and those due to unreal navigational VBPs.

9.3.1   Visual-vestibular (or visual-proprioceptive) discrepancy

This is the classic case of virtual movement by vection without the observer actually moving. This discrepancy is disturbing to the user as soon as it exceeds certain limits of cinematic movement. Movement can be controlled manually, via the head or feet, and in the latter case will require a standard treadmill (1D) or an omnidirectional treadmill (2D)[3]. As indicated previously, this case is more natural and consistent with virtual motion via vection, as there will be coherence between the vision and the proprioceptive organs in the muscles, tendons and joints. However, the vestibular systems are poorly stimulated, as the person remains motionless in the RE.

Reducing the discrepancies (S15, S16, S17, S18, S19 and S20)

There are six different solutions; the first two are complementary as are the last four:

–   In order to not overtax the vestibular systems, translational and rotational accelerations should be limited, as should be tilted movements of the virtual camera (the viewpoint of the person in the VE) filming the virtual scene (S15).

---

[3]There are also other foot-controlled devices.

- To avoid overtaxing the vestibular systems, the design of the VE must ensure that the trajectories of the virtual camera are not excessively convoluted; consequently, they should have high radii of curvature, which will limit the centripetal accelerations that are a normal feature of the trajectory (S16).
- The perception of movement is more sensitive at the periphery of the visual field, detecting optical flows due to vection movements and to movements of objects in the scene. Reducing the observed field of view may be envisaged by hiding images in the peripheral vision, even if most of the VR headsets currently offer a fairly restricted field of view. This solution is at the expense of the visual immersion experience (S17).
- As previously indicated, when users look at the VE on an ordinary flat screen – and not via a VR headset – their peripheral vision when looking at the VE, even unconsciously, would stabilise the images and prevent them from feeling discomforted or unwell. To get closer to this situation, it is possible to reduce the discrepancy by injecting into the images on the peripheral vision some of the spatial references of the real environment for the purpose of stabilising the user. This will reduce the optical flows in the peripheral vision. Naturally, the VR headset must have a fairly broad field of view, at least 100°, to ensure it does not negatively affect the central vision. These artificial and unrealistic inclusions can disrupt the visual perception of the observer although, in principle, fairly moderately as they are located in the peripheral vision (S18).
- Following on from the previous solution, it is possible to minimise the discrepancy by having objects in the VE that are motionless relative to the RE, when the user is also motionless in the RE. The classic example to illustrate this point is that of a static driving simulator: the driver feels adequately stabilised if he can see the interior of the car in his peripheral vision, when he is actually motionless in the RE (S19).
- To return to the previous solution, it may be worth using a VR headset with a visor that does not totally obscure the vision ("video-goggles") to allow the observer to directly perceive the real world in his/her peripheral vision. Under these conditions, the disruptive effects of the visual-vestibular discrepancy are greatly diminished, and will be fairly similar to looking at an ordinary screen (S20).

We will now provide details of S15, S16, S17 and S18.

Details of solution S15

It is obvious that to avoid overtaxing the vestibular systems, no translational and rotational acceleration and no tilting of the person's head must occur, given their anatomic constitution and functionality. Conversely, if the intention is to place demands on the vestibular systems, we will need to act on the accelerations and head tilting. This imposes for transport simulations as well as VR applications that actions must be taken on the person's body with motion simulation interfaces – a solution explained below (S24). This solution requires a fairly substantial investment in terms of equipment. In the classical case without adding a behavioural interface, the choice involves restricting the accelerations and head tilting to limit the visual-vestibular discrepancy. These limits cannot be determined absolutely as they depend on several parameters: duration of incoherent demands and the capabilities of each individual, as well as on the retinal disparity limitations for stereoscopic vision.

Rules to limit the disruptions of the visual-vestibular discrepancy have already been extensively studied in static transport simulators (without motion simulation interface). The assumption is that these rules may also apply to VR applications with headsets, when all the basic settings are identical: same field of view, same type of movement trajectory, which limits the correspondence between the two situations. Most of the time the display will differ, as simulators generally use large screens. However, the transport-simulation and virtual-reality specialists at the Renault car manufacturer have also used VR headsets for transport simulation (Kemeny, 2014).

It is recommended that movement speeds are set to a usual walking pace (3 to 5 km/h) or, if necessary, a running pace (10 to 12 km/h) depending on the context. It can be assumed that speeds, with an influence on the vestibular-ocular reflex[4], especially detected by the peripheral vision (when the VR headset has broad field of view) affect the simulation sickness, even if accelerations are essential. Consequently, it is not the intensity of the speed that plays an important role in the simulation sickness, but it is the intensity of the acceleration and even variations in acceleration: that is, jerking movements that correspond to sudden jolts. The duration of the accelerations has an influence: the accelerations should only last for a short time, which is understandable, as the discrepancy will therefore last for a shorter time. The programming should be adapted as far as possible to limit the duration of any acceleration and ensure acceleration only occurs infrequently (Stanney, 1998).

If it the users who freely and totally control their movements, it is not possible to place constraints upon them. However, by partially controlling the movements of users, it is possible to limit their movement-control signals, thereby ensuring they are unable to generate excessive accelerations. For example, the relationship between the control signal and the acceleration value, rather than being linear, is non-linear and limited asymptotically. It is also possible to impose the shape of curvature of the trajectory controlled by the user by imposing radii of curvature that are sufficiently large to limit high accelerations (see solution S16) through the appropriate controls (see above).

In any event, it is preferable that the accelerations are imposed on users, as they will be able to anticipate their reactions and will be therefore less susceptible to simulation sickness. Acceptable acceleration values are around 2 m/s^2 for a translation movement for a few seconds and 2°/s^2 for a rotation movement (Kemeny, 2015). However, the acceleration for a 13°/s^2 rotation movement will be disruptive for novice users and is barely acceptable (see chapter 11 for industrial applications). Limiting rapid acceleration changes (sudden shocks or jerks), which most people find hard to tolerate, is also recommended. For example, rapidly descending a staircase should not appear too realistic (by using a motion vection), and jerking movements should not occur on each individual stair. It is preferable that the trajectory should be purely rectilinear at a constant speed.

We have stated that the brain anticipates the future sensory stimuli it receives. When users control their movements, the brain is able to anticipate; however, when

---

[4]As stated above, the vestibulo-ocular reflex makes the eye swivel in the opposite direction to the head's movement. Vestibular systems, by measuring the head's rotation, order the eyes to move in a direction opposite to the movement of the head to stabilise the retinal images.

users are passive relative to their movements and trajectory, it is a good idea to provide sensory cues to help anticipate any changes in acceleration. In real-life vehicles, and to prevent motion sickness, noises can be added to the gearbox; therefore, passengers who are unable to see the driver shifting the gears can anticipate accelerations of the vehicle. It could be worthwhile to use the same approach in virtual reality, as it would help users anticipate their movements.

When moving, users can simultaneously observe the VE and therefore turn their heads. Coupling two of the VBPs (navigation and observation) will place even greater demands on the users. When users control their own movements, we recommend that they should be able to manually control their movements while simultaneously being able to look around in all directions via the head controls. This is navigation and observation in "tank mode", which is interesting to use[5]. However, we are accustomed to look 'ahead' or slightly sideways. Consequently, the angle between the direction of travel and VE viewpoint angle should therefore be limited. Depending on the user's abilities, this angle can be limited to a greater or lesser extent by programming or by encouraging the user to always look ahead of themselves, or by inducing this attitude through the scenario of the video game or the visual or audio elements of the VE. When users are sitting on a chair, they will not tend to turn their heads beyond a $-45°$ to $+45°$ area. This can be taken into account when programming movements (by vection) so that they retain directions with low angles in relation to the user's viewpoint axis (and therefore no "virtual/real walking backwards", i.e. when the direction of the virtual travel is the reverse direction to that of the real-life seat). If necessary, a resetting option must be available if the angle between the direction of travel and the VE viewpoint angle is too high. Another method of travel is one using a constantly zero angle: the direction of the virtual travel results from the direction of the user's head, which is less problematic as this type of navigation is similar to walking in real life. Moreover, the travel can be totally free or, conversely, depend on the trajectory determined by the designer of the VR application. In this case, the trajectory will influence the accelerations experienced by the person, even if it is he or she who controls the speed of travel over the pre-set trajectory. This issue is presented in the following paragraph.

Details of solution S16

This solution describes situations in which the trajectory is determined by the designer of the VR application, regardless of whether or not the person is controlling their own movements. Any predefined trajectory will influence the accelerations experienced by the person whenever the trajectory is not straight. Reminder: in a circular movement, centripetal acceleration, which is a normal part of any trajectory, is inversely proportional to the radius of curvature and is proportional to the square of the tangential velocity. Speed in a curved trajectory must therefore be limited. When going around a corner in a car, the tighter the bend then the smaller the radius of curvature will be, and accordingly the greater the acceleration experienced. When the trajectory is neither flat nor on a 2D surface but is in a 3D space, the problem of limiting accelerations is even

---

[5] https://developer3.oculus.com/documentation/intro-vr/latest/concepts/bp_app_ui_nav/

harder to manage. Rotational acceleration occurs whenever there is a bend. To min-imise rotational accelerations, trajectories with relatively large radii of curvature radii are required to avoid jolts and sudden changes in acceleration, and it is preferable to have a radius of curvature that is established gradually. Designing trajectories shaped to avoid jolts has been a problem for over a century, since the construction of the first railways, followed by roads and motorways – the issue concerns the clothoid arcs[6]. Application designers therefore need to create trajectories that are imposed upon users in an identical fashion, whether these are 2D or 3D trajectories. This issue has been partially studied for roads with bends of a certain slant to enhance the comfort of passengers in vehicles. A relationship has been established between the slope of the slant, the radius of curvature and speed, to maximise comfort. These same concerns must also be transferred to virtual reality in order to design "soft" trajectories, that is to say ones that minimise the "visual shaking" of the person.

When the trajectories depend on the geometry of the VE, this geometry must be analysed to avoid as far as possible rotational movements and, more generally, untimely accelerations. Care must be taken regarding the layout of obstacles in the VE. For example, in a virtual store, seen via immersion on a large screen located very close to the observer (as would be the case with a VR headset), the layout of aisles and shelves is designed to ensure that the person can move virtually without having to rotate. With a VR headset, this becomes more complicated if, at the same time, as the person is moving, he/she turns their head, as we have already mentioned. The geometry of the VE which imposes specific and predetermined trajectories should be thoroughly analysed to minimise the visual-vestibular discrepancy. Trajectories may be restricted partially but not totally constrained geometrically – for instance when walking down a corridor. This means that users will have a certain leeway to move about even though the corridor leads only to one place. In this situation, users may feel that they are totally free to direct their own movements. We have already tested this type of situation without the knowledge of users who were surprised after learning that their trajectory had been restricted.

If realistic movements are too restrictive, nothing prevents an unreal VBPs (VBP Ur) from being designed. This is particularly the case when you want the user to turn around, as the operation can be extremely disturbing in VR, if you need to make it real-istic. It may be preferable to use a metaphor and not a schema, i.e. using the metaphor of the "inverted tunnel": when a user moves forward, on the side of the trajectory there is always a symbolic entrance to a tunnel. And, by heading towards the entrance, the user rapidly enters and exits the tunnel – in just one or two seconds – but will be on the same trajectory he/she has just left, but travelling in the opposite direction (Figure 9.7). This means that the person will have turned around on him- or herself via a vection mechanism. Guidance cues should be provided in the VE to prevent users from becoming disoriented (sunlight, spatially located sound source, different buildings on each side of a road, a range of plants and types of nature, etc.). Other

---

[6]In 1890, the engineer Talbot created equations to determine the shape of railway tracks at the start of a bend to eliminate jerking movements when accelerating: namely, clothoid arcs, whose radius of curvature changes gradually. For the same reason, a driver will gradually turn the car's steering wheel at the start of a bend.

Figure 9.7 It may be preferable to use a metaphor and not a schema: the "inverted tunnel".

unrealistic travel metaphors may also be envisaged to rotate one quarter of a turn, for lateral time-lags and instant ascents, etc.

In the real world, we are almost always moving forward and rarely sidewards. Consequently, forward movements are more realistic and less disturbing. However, the virtual space created does not need to offer a topography that is realistic and geometrically identical to that of the real world. The topology of this space can be discontinuous, anisotropic, not fixed but changeable over time, not necessarily indivisible or may be illogical, etc. And this can help to avoid or minimise the visual-vestibular discrepancy, as was the case with the metaphor of the inverted tunnel. Imagination is needed to devise unrealistic but effective solutions in terms of comfort and usage.

To conclude, it may be worth setting a "budget for accelerations" in order to limit translation and rotation movements, depending on the intensity and duration of the accelerations, applying a similar procedure to the idea of having a "budget for stereoscopic vision".

Details of solution S17

The perception of movement is more sensitive at the periphery of the visual field. The field of view observed can be reduced by concealing part of the peripheral vision. The basic idea is to design one of the areas of the peripheral vision shaded black. However, it could make more sense to use the average colour of the scene in this area (or the average colour of one section of the backdrop) to limit less the immersive perception in the VE. The aim is to limit the discrepancy between the vestibular systems and the vection created by the optical flows triggered by the person's movements and the movement of objects in the VE. Reducing the field of view reduces the discrepancy. However, it may be wiser to only conceal the optical flows linked to the person's movements (viewpoint) without concealing in the peripheral vision moving objects relative to the VE, in order to better understand the scene and thereby improve the immersive experience. A test[7] was performed by only eliminating part of the peripheral vision – the section corresponding approximately to concealing the nose[8]. This partly alleviated the sensation of feeling unwell. This amounts to applying the solution described above, making the person believe that the pseudo-concealment of the nose improved their stability; however, and according to specialists, concealing the nose does not have any impact on stability in the RE.

Details of solution S18

It is possible to reduce the discrepancy by injecting into the images on the peripheral vision, some of the spatial references of the RE, to act as a landmark and stabilise users. This will reduce the optical flows in the peripheral vision. Naturally, the VR headset must have a fairly broad field of view, at least 100°, to ensure it does not negatively affect the central vision. The head-tracking system also needs to be accurate and correctly calibrated in relation to the RE, to add landmarks from the real world that will remain motionless in relation to the RE when the observer moves his head. This requires a 6 DOF head tracker sensor and, preferably, a 3D reconstruction device for the real environment to properly reposition the person and his headset in the RE. It would be technically interesting to attach the real-environment 3D reconstruction device to the VR headset. These artificial and unrealistic inclusions can disrupt the visual perception of the observer although, in principle, fairly moderately as they are located in the peripheral vision and the user will not necessarily be aware of them. The number of landmarks will need to be adjusted, given that they occupy a large section of the peripheral vision, which is somewhat reminiscent of the previous solution (S17). These landmarks can represent the shapes of real objects (furniture in a room, etc.) in the RE (Figure 9.8). When the user turns his head, he will see these landmarks move, such as, for instance, straight lines in his peripheral vision and in his natural vision in the RE. This may seem strange but it can help minimise the visual-vestibular discrepancies. These landmarks do not need to be permanently displayed but only during rotational movements or translational accelerations. These landmarks can take different forms: segments, set of points, partial grids, etc.

One interesting use, with a reconstruction of the actual place (the RE) in 3D point cloud via Kinect, was conducted by David Nahon and his team of the iV Lab at Dassault Systèmes. The[9] "Never Blind in VR" experience provides an overview of the real environment to users of a VR headset in the form of a cloud of 3D points corresponding to surfaces detected by the Kinect. This solution offers several benefits: it allows you to perceive your own body in the real environment and other people in real time and correctly co-localised, when it has been correctly calibrated. This solution

---

[7]http://phys.org/news/2015-03-virtual-nose-simulator-sickness-video.html
[8]The virtual concealment of the lower (and inner) section of the right-hand and left-side images does not really correspond to an attempt to conceal the real nose, as the field of view in the VR headset is obviously much smaller than the real field of view.
[9]http://www.3ds.com/fr/recits/never-blind-in-vr/

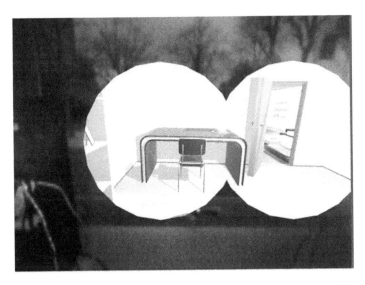

Figure 9.8 To reduce the visual-vestibular discrepancy, we tested the S18 solution and obtained positive results when moving in the VE we embedded in the peripheral vision a view of the RE taken using a Leap Motion device attached to the VR headset.

allows actions to be taken in the VE with the same level of safety as that offered in the RE (see below). By displaying points at the level of peripheral vision at all times, the person may feel more stable despite the visual-vestibular discrepancy.

The effectiveness of these solutions has yet to be validated scientifically. They may improve when VR headsets have integrated eye-tracking to be more effective.

Removing the discrepancy by changing the way VBPs operate (S21, S22 and S23)

There are three different solutions:

– The actual movements of the person standing in the RE, is the geometrically identical to movements in the VE. Trajectories and speeds in the VE and RE are always the same. In this case, both the real and virtual environments must have the same dimensions that is they must be **geometrically identical** (Figure 9.9). For example: if the real environment is a room with standard dimensions in which the user is standing and can move about, any movement in the virtual environment must be identical and limited by the room's dimensions. This implies a coherence between the visual stimuli, those of vestibular systems but also those of other proprioceptive stimuli (neuromuscular spindles, Golgi tendon organs and joint receptors), and identical gestures both in the RE and the VE. The visual representation of the VE generally includes a co-localised avatar (or at least co-localised hands, if the user does not need to look at his entire body) to help the user better immerse himself bodily and easily manipulate objects, thanks to the co-localisation of his hands (**S21**);

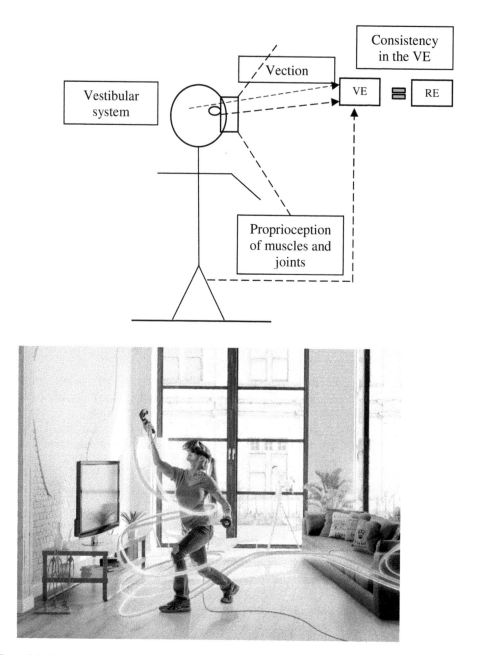

Figure 9.9 Both the real and virtual environments have the same dimensions and are geometrically identical. Photo: a user with a HTC Vive headset (courtesy of HTC & Valve).

- Travel in the VE is carried out via **teleportation** from one place to another in the VE, while remaining motionless in the RE. Thus, continuous movements are eliminated and no demands are placed on the vestibular systems since there is no longer any speed or acceleration. The two senses are therefore coherent as the person is motionless both in the RE and the VE. The place of arrival is generally selected using a localisation sensor with one hand, indicating the point of arrival virtually (metaphorical process). The place of arrival must be visible from the place of departure, otherwise you will need to point to a map of the location. The user goes virtually from the place of departure to the place of arrival instantaneously. However, the visual transition from the point of view of the place of departure to the place of arrival may be achieved using a fade out to soften the teleportation (S22).

- A radical solution is to produce the application in **Augmented Reality**, which then imposes the condition detailed in solution S21: both the real and the virtual environments must obviously be geometrically identical, because they are merged, and any discrepancy is therefore eliminated. On a technical level, this requires using an Augmented Reality headset. The user sees the real world, meaning he is even more stabilised thanks to his peripheral vision which lies in the RE, as technically it is not possible to display peripheral-vision images in AR headsets; however, the options offered by AR are not the same as those offered by VR. Each area, with very different usages, has its own benefits and limitations (S23).

Removing the discrepancy by adding/modifying behavioural interfaces (S24 and S25)

We can eliminate this discrepancy by physically creating correct stimuli for the vestibular systems using a motion simulation interface or a control interface via a 1D or 2D treadmill. Acceleration and gradient stimuli on the person's body must, if possible, constantly match the person's visual movements (vection):

- With a motion simulation interface, demands are not only placed on the vestibular systems but also on the proprioceptive organs of muscles, tendons and joints, which must also be coherent. Depending on the virtual movement required, the motion simulation interface must be able to create the correct stimuli for the vestibular systems, given that in some cases, there is no perfect solution, even without taking the price of the interface into account. For example, for the car-driving simulation, the most high-performance simulators from the car manufacturers, costing several million euros, are able to create correctly most driving simulations, with the exception of certain trajectories, such as ones around a roundabout where the actual accelerations suffered by the driver's body are too high (S24);

- With a 1D or 2D walking treadmill interface and other similar devices, the proprioceptive organs of the muscles, tendons and joint are able to be correctly simulated to reduce the discrepancies, while this is not the case for vestibular systems. The visual-proprioceptive discrepancy will be reduced in this case (overall proprioception) but the visual-vestibular discrepancy will always be present (Figure 9.10). One of the solutions from S15 to S20 should be taken into account and used to minimise the visual-vestibular discrepancy (S25).

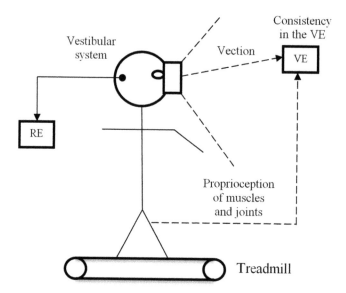

Figure 9.10 With a treadmill, the proprioceptive organs of the muscles, tendons and joints are correctly simulated in relation to the vision; however, the visual-vestibular discrepancy is still present.

Details of solution S24

Several types of motion simulation interfaces can be used depending on the movement required; however, there is a need to determine the one most suited to triggering the correct stimuli for the vestibular systems. At times, the exorbitant cost of these types of interface means that a less effective though more affordable one is used. For general-public applications, six-axis mobile platforms are reserved for use in amusement parks. Mobile seats with a few degrees of freedom (2 or 3 in general) can also be used but their price means that they cannot be used at home, only in games rooms. Vibrating seats can make some more simulations more attractive. In general, however, the interaction will be manual and in addition to the feet. Users will therefore have the option to observe the VE independently while they are moving. The same precautions as those applicable to solution S15 should be taken when a visual-vestibular discrepancy is present due to a motion simulation interface that is not totally appropriate for the movements in the VE.

Details of solution S25

With a walking interface, such as 1D or 2D treadmills or similar devices (rowing machine, exercise bike, etc.), one of the solutions from S15 to S20 should be used. For 1D treadmills, a manually operated system for turning must be available. Attention should also be paid to the risk of becoming disoriented, as is the case with other methods of virtual motion. The advantage of a treadmill is that users are able to better perceive distances, which can be useful for some VR applications, and to become more immersed on a sensory level. The downside, however, is that users may be apprehensive, as they are unable to see their own feet, or at least not directly. It is preferable that users should be able to perceive a correctly co-localised representation of their feet or, alternatively, bars should surround the users to help them keep their balance.

With such devices, the benefit of using a VR headset rather than one or more screen surrounding the person's head is debatable. It is amazing to see a rowing simulator application with a home rower and a VR headset. Rather than using a VR headset, it would be better to use two or three flat and fixed screens, placed in a U-shape, and surrounding the viewer's head, as explained in solution S3. The images will be of much better quality and calculated irrespective of the position of the head, which moves hardly at all in relation to the landscape seen from a boat. If this is the case, the head's rotation will no longer need to be measured, as the discrepancy due to latency is eliminated. However, the latency lag between the movement command via the oars and the correct display of images with the correct viewpoint will still remain. But as the movement is slow and the scenery is relatively far away, the latency will not be perceptible with this virtual rowing application.

To complete this section on visual-vestibular discrepancy, it may also be worth considering stabilising the person using spatial and fixed sound sources in the RE, even though their location is incoherent relative to the sounds required in the VE.

9.3.2 Temporal visual-vestibular discrepancy

The latency lag between the movements of the mobile platform and displaying the correct viewpoint should be minimal to avoid any temporal visual-vestibular discrepancies.

Reducing the discrepancy (S26)

The latency should be reduced technically and using computer resources.

Removing the discrepancy by changing the way VBPs operate

Inappropriate request.

Removing the discrepancy by adding/modifying behavioural interfaces (S27)

To eliminate the latency between the mobile platform's movements and displaying the correct viewpoint, simultaneous controls are required to synchronise the platform's movements with the images displayed. For this synchronization, the latency of the head *tracking*, the motor control interface (e.g., the steering wheel in a transport simulator) and the movement of the motion simulation interface all need to be taken into account. However, and even after synchronization, the latency lag between the navigational controls and the two synchronized sensory simulations will still be present.

9.3.3 Visual-postural discrepancy

This discrepancy may occur when the movement in the VE takes place via vection when the user is standing and is motionless – or almost motionless – in the RE. The central nervous system receives information from the vestibular systems and also from other proprioceptive stimuli (muscle spindles, Golgi tendon organs and joint receptors) about the fixed position of the user's body.

Reducing the discrepancy (S28)

We can minimise this discrepancy by intervening on the virtual movement (type of trajectory, movement kinematics) using the same solutions as those used for the visual-vestibular discrepancy (visual-proprioceptive). It may be useful to represent the users' body (their avatar) in the VE, to help them situate themselves in the VE in relation to their proprioceptive perception in the RE. When users are able to see their co-localised feet virtually in the VE, they will perceive a coherent view of their height, eye level and therefore the position of their viewpoint in the VE in relation to the tactile stimuli under their feet in the RE. When this is not the case, users can feel uneasy and will not feel stable.

Removing the discrepancy by changing the way VBPs operate (S29)

We get users to sit down or make them hold on to a support for greater stability when they are in a standing position, to provide greater reassurance.

Removing the discrepancy by adding/modifying behavioural interfaces
Inappropriate request.

9.4 MANIPULATION VBPS

To simplify the explanations, in the following paragraphs we will not differentiate between discrepancies due to the technical failings of the VR headsets and those due to unreal navigational VBPs.

9.4.1 Visual-manual discrepancy

In terms of manipulating objects, etc. in the VE, if the real-life hand of the user is poorly co-localised in relation to the view of the hand he or she has via the VR headset, this will trigger a hand-eye coordination discrepancy and the user will have to adjust to this situation accordingly. This will not make users feel unwell but it may cause discomfort if the action is hard to control. The hand does not always need to be represented. A device (such as Leap Motion) that detects in real time the movements of the fingers allows for more realistic simulations and more delicate manipulation controls. However, in the VE, one of the advantages is that you do not have to use the fingers to perform the manipulations, as fairly unreal manipulation VBP or BSAs (Behavioural Software Aids) can be used for this purpose.

Reducing the discrepancy (S30)

We reduce the bias between the proprioception of the user's actual hand and the vision of their virtual hand in the VR headset by improving the accuracy of the sensors and calibrating correctly the various landmarks: the viewpoint landmark in the VR headset and the two position landmarks of the two head-tracking and hand-tracking sensors.

Removing the discrepancy by changing the way VBPs operate (S31)

It is always possible, instead of using the sensorimotor schema for manual manipulation tasks, to manipulate an object via a metaphorical action. A wide range of options are available: manipulating with a joystick, clicking on an object with a virtual laser and moving by indicating the point of arrival, manipulating with a non co-localised hand via teleoperation, etc.

Removing the discrepancy by adding/modifying behavioural interfaces (S32)

Manual manipulations are possible with a tactile feedback interface and/or with a force feedback interface in order to receive coherent stimuli between the touch, proprioception and vision. In general, with a force feedback interface, the hand's position is measured precisely by the interface, which therefore eliminates any visual-manual discrepancy. Conversely, when a force feedback interface is not used, a sensorimotor discrepancy will occur when one object collides with another: no contact force and no blockage of the hand's movement. A BSA (Behavioural Software Aid) can then be used by adding specific constraints applicable to the manipulation of a 6 degrees-of-freedom (6DOF) object, to help the user to achieve the desired motivity (e.g. placing an object on a table when the movements of the virtual object are restricted). The support surface of the object should be parallel to the table's surface when they are close to one another. These unreal constraints are called virtual (movement) guides or magnetism effects.

9.5 ANALYSIS GRID FOR THE 32 SOLUTIONS

On the basis of the list of 11 disruptive sensorimotor discrepancies, the following three questions have been answered:

– How can you minimise the impact of sensorimotor discrepancies in terms of the discomfort of users or the sensation of feeling unwell?
– It is possible to eliminate the sensorimotor discrepancies by modifying the way VBPs work?
– Can we eliminate sensorimotor discrepancies by changing the way the interface works or by adding another interface?

These questions have allowed us to describe 32 solutions. Some of the solutions are dependent on the future development of certain features on the VR headsets; others require changes to the equipment (interface) while others relate to the functional design (VBPs) of the VR application. Some of the solutions are highly specific and will only be used very rarely. It is always possible during a VR application to switch from one solution to another on the condition, naturally, that the user is able to adapt to the change to the virtual behaviour primitives. The final question to be analysed in the next section concerns the way we can adapt these discrepancies to ensure that VR users do not suffer any discomfort or feel unwell.

To summarise, solutions are available to:

- change the domain: from virtual reality to augmented reality;
- change the visual interface, when the VR headset is not the optimal solution;
- change the VR headset to use a more high-performance product;
- change or add an interface associated with the VR headset;
- change an unreal VBP for a conventional VBP;
- modify the operation of one of the VBPs;
- remove a disruptive sensorimotor discrepancy;
- minimise the effects of the disruptive sensorimotor discrepancy.

However, it should not be forgotten that users can also adjust to a disruptive sensorimotor discrepancy (see below). Finally, it is sometimes necessary to wait for techniques to evolve in order to use a future VR headset that offers the appropriate features, if the previous solutions are unsatisfactory for the VR application required.

Designers of VR applications, with the exception of solutions that depend on the manufacturers of VR headsets, must be aware of the solutions applicable to the three main disruptive sensorimotor discrepancies that are both the most important to manage and the best known (without overlooking other solutions for certain types of VR applications):

- temporal visual-motor discrepancy (head tracking latency);
- oculomotor discrepancy;
- visual-vestibular discrepancy.

The solutions for these three discrepancies are summarised in the following table:

Table 9.1 Summary of solutions to consider for the management of the three main sensorimotor discrepancies.

	Minimise the impact of a sensorimotor discrepancy	Change the operation of a VBP	Remove a sensorimotor discrepancy
Temporal visual-motor discrepancy Oculomotor discrepancy	S1: reduce software execution times S6: reduce retinal disparities and image processing	S2: sequence of frozen images S7: monoscopic vision	S3: screens surrounding the head without translation S8: Light Field or VRD screens
Visual-vestibular discrepancy	S15: limit accelerations and inclines S16: soft trajectories S17: reduce the field of view S18: processing in the peripheral vision S19: fixed objects/RE in peripheral vision S20: peripheral vision not concealed	S21: RE and VE geometrically identical S22: travel via teleportation S23: Augmented Reality application	S24: with a motion simulation interface S25: with a 1D or 2D treadmill

The 12 most important solutions are indicated by the letter S in bold. The secondary solutions are indicated in italics and suggest either:

– changing the domain (AR rather than VR) and therefore the type of VR headset;
– changing the visual interface;
– adopting highly specific solutions.

Technical and financial considerations

Some of the solutions are expensive

These are barely feasible for **individual purchases of video games for personal use**, unless targeted at persons with considerable financial resources, thereby implying a niche market for such solutions. The most expensive of these are reserved for amusement parks (S24) or video gaming rooms (S25), even if companies are attempting to sell treadmills for less than one thousand euros.

Solution S21 (RE and VE geometrically identical) requires a room dedicated to the video game (a "scaleroom", measuring approximately $10 \, m^2$), unless the spatial activity of the game is limited by making the player remained seated. The advantage of having a dedicated room is that it can be fitted with permanent tracking cameras, to avoid having to constantly calibrate the *tracking*. This is the technical solution proposed by the major manufacturers of VR headsets, as it helps solve the very low latency (S1) problems and offers high accuracy (S11) for tracking, as long as the developer does not increase the latencies. This solution (S21) remains fairly expensive and bulky, as it requires a small dedicated room or space in a home. And if the use of a larger space than a room is required, solution no. 21 is reserved for gaming rooms or amusement parks that can use a large hall ($400 \, m^2$, for example[10]).

As these are **professional VR applications**, the financial constraints are not the same, which paves the way to the use of a greater number of solutions – with or without the use of a VR headset for the visual interfacing – some of which have already been used for around the last twenty years.

The nine other main solutions (S1, S6, S7, S15, S16, S17, S18, S19 and S22) are under the control of the designer of the VR application and do not involve any additional financial cost, beyond the costs arising from potential new developments.

Solutions limiting the VBPs

Solutions S7 (monoscopic vision) and S6 (well-controlled stereoscopic vision) do not involve any major constraints, and only require settings adjusted to the individual capabilities of each headset wearer. Three-dimensional perception using stereopsis in the VE is not essential in very many applications. It can be removed (S7) or highly minimised (S6); however, visual immersion is less attractive and delicate manipulation of objects located close by is not as easy. Similarly, solution S17 (limitation of the field of view) is easy to apply but decreases the peripheral vision of the observer, which in some cases can be offset by spatial sound or other visual tricks. These three solutions limit the observational VBPs and make it less attractive. Solution S22 (teleportation) impose navigational VBPs that are less attractive but are just as effective.

---

[10]See https//www. zerolatencyvr.com

Solutions controlled by the designer

Regarding the five remaining solutions (S1, S15, S16, S18 and S19):

- Solution S19 requires at least one fixed virtual object compared with the RE and therefore imposes a certain type of VE and video-game scenario (car cockpit, etc.).
- Solutions S15, S16, S18 are under the control of the designer and are used to limit the negative effects of the visual-vestibular discrepancy. Solution S18 may, in one of its subcategories, and at an additional cost, require a 3D reconstruction of the RE.
- Solution S1 (very low latency) imposes on application developers the requirement to not increase the latency. Developers must create programmes with a phase to optimise the functions used and offer a software architecture favouring parallelisation, to minimise the overall latency.

All designers or developers are responsible for setting the parameters (acceptable thresholds for the solution, etc.). This is impossible to achieve without ascertaining the level of adjustment of any individual to the VE, which is the purpose of the next section.

9.6 ADAPTING TO THE VIRTUAL ENVIRONMENT

9.6.1 Levels of adjustment difficulties

The issue of any person's adaptation to his/her immersion and interaction in a virtual environment is more extensive than the mere adjustment to sensorimotor discrepancies to avoid discomfort and feeling ill. In connection with our $3I^2$ interfacing model, four points will be examined, with the first two relating to the I^2 sensorimotor level:

- physiological adjustment to the visual interface: the VR headset;
- adjustment to sensorimotor discrepancies;
- cognitive adjustment to interfacing;
- functional adjustment to the VBPs.

The physiological adjustment to the VR headset requires adjusting correctly the headset and making sure the interfacing with the visual system of the user is appropriate, given that optical capabilities vary considerably from one person to another. It should not be overlooked that before each VR application or, at least, before using a new VR headset for the first time, ergonomic and optical calibrations and evaluation tests of the visual interface must be carried out. If these operations are not carried out or are insufficient, the recommended solutions for adjusting to the sensorimotor discrepancies may not be sufficiently effective. For example, it is difficult to reduce the "accommodation-vergence" discrepancy in relation to the stereoscopic vision when the user has not correctly interfaced the headset on an optical level, especially when the person in question has an eyesight deficiency. We must not forget that in a VR headset we do not exactly know what the observer is seeing.

Cognitive adjustment to the interfacing should be checked to ensure that the person has understood the cognitive interfacing process and is able to control it in practice, whether it involves an Imported Behavioural Schema or a metaphor. Some failures occur as the cognitive process used by the user was not the one envisaged by the designer of the VR application.

The functional adjustment to the three VBPs must be also evaluated. For people who are more sensitive to sensorimotor discrepancies it is highly recommended to refrain from using unreal Virtual Behavioural Primitives (VBP Ur), and that these should only be reserved for users with a high adaptive capacity.

In VR, the main difficulty lies in getting people to adapt to the sensorimotor discrepancies to avoid discomfort and the sensation of feeling ill and, in the worst-case scenario, the rejection of the VR application. An **Overall Level of Adaptation Difficulties (OLAD)** should be established based on the list of 11 discrepancies and the solutions used. Several solutions can be proposed in a VR application according to the adaptation difficulties, which involves several configurations of the solutions to be programmed. Some of the solutions have variable parameters (threshold value, etc.). They must also intervene in the determination of the OLADs and the configuration of the settings. We can quantify the OLAD for each of the configuration of the application. Each user can choose his/her configuration according to their sensitivity to sensorimotor discrepancies.

Research should be carried out on the impact on the health and safety of users on the basis of the eleven discrepancies and the solutions envisaged. The severity and extent of the impact will change over time according to the latest technical developments and the solutions implemented by the designer of the VR application. Designers should manage the intensity of the impacts and tailor the exposure time to each individual, which implies determining a budget. Video games could be programmed to a stop, or at least require users to take a break after a certain period of time, independently from the wishes of the player; this is the situation with 3D cinema, as spectators do not chose the length of the three-dimensional film they are watching.

Adjustment tests could be offered to allow each user to determine the impact of the various discrepancies on an individual level. Learning exercises could also be supplied. Users could apply some of the solutions and set the parameters themselves, for example: switching to monoscopic vision or lessening the impact of stereoscopic vision using a personal setting to determine the maximum limit of retinal disparity.

9.6.2 Levels of user adaptation

Adapting to sensorimotor discrepancies will depend on each user. Based on experience and some classic and well-analysed cases we are well aware that almost all people are able to overcome some discrepancies by adjusting to them. Designers of VR applications should not forget to take into account the person's morphology namely the interpupillary distance (IPD), the height of their eyes from the ground, etc.

The age of the users can be taken into account though it should be remembered that no causal relationship has been determined between age and the ability to adapt, since other characteristics are involved such as the experience of users with VR applications, for instance. This characteristic is a positive factor which allows users to be less sensitive to discrepancies, as the brain has a well-recognised ability to adapt (hypothesis 9). This is also true for video game designers who gradually adapt to their new applications during their own evaluation testing. They must therefore get novice users to test their applications. And they must start the tests with the lowest level of OLAD.

However, the adaptive ability will also depend on the motivation of the player. Some sequences in the VR applications can be offered with major sensorimotor discrepancies, when players are motivated and wish to have an "adrenaline-fuelled" experience. Such discrepancies will then more easily accepted, as they are in thrilling "roller-coaster type" attractions.

9.7 SAFETY RULES

The main safety problem is due to the visual and acoustic insulation of the person wearing the VR headset, with the obvious exception of AR headsets. How can we overcome the lack of direct visibility of the real environment where the person is actually located, especially if the layout of the RE changes during the course of the application? It is particularly important to take the physical safety of the person into account, especially when the person is standing in a room rather than sitting down. In the latter case, the visual and acoustic insulation of the person prevents them from being alerted to any outside event unrelated to the activity in question.

On the other hand, if the person is standing in a room, the physical activity related to the VR application can be a source of danger, especially if movement is required, even it the movement only involves moving around one or two metres. In this mode, the minimum level of safety required is that the person should receive an alarm whenever he or she approaches the walls of the room or the furniture. It is highly desirable, not to say essential, that the head-tracking sensor on the VR headset is 6 DOF and that the VR application is aware of the positions of the walls of the room, or an area without furniture. Information on the relative positions of the walls in relation to the VR headset can make the movements of the person safer. When the person comes too near the walls (or furniture), an alarm will provide a warning that there is a risk of collision. It can be a sound warning or even a touch signal; the alarm can also be given visually by the means of a virtual grid or another visual clue indicating a limit that should not be crossed. An even more effective safety measure would be to suppress images or have the application stop. It could be possible to combine the different types of alarms simultaneously or slightly stagger them to provide progressive warnings. Regardless of having a safety system, the VR application should preferably not be used in areas with specific potential hazards such as a kitchen, bathroom, room with a balcony or providing access to stairs, etc.

The relative positions of the walls in relation to the VR headset can be established prior to the start of the VR application at the time of its configuration; however, it is better that this stage is executed in real time. This obviously depends on the head-tracking method. For example, if it requires cameras positioned at the four corners of the ceiling of the room, the free area of movement without danger can be defined in the head-tracking configuration. If the VR headset includes an integrated 3D reconstruction device determining in real time the position of the walls, furniture and persons present (i.e. kinect from Microsoft or a frontal camera such as the one on the VIVE headset from HTC), safety will be correctly assured. The "Never Blind in VR" application from Lab iV at Dassault Systèmes ensures the safety of the VR application, as the computer permanently knows the location of the user and what is happening around him. In all situations, it is preferable that the hands and, better still, the entire the body are represented virtually in the VR headset, which poses no technical problem with devices that offer a real-time depth map of the site.

If the person is sitting on a mobile platform, special safety conditions must be provided, which are normally integrated into the structure, such as safety belts, if the platform places considerable demands on the user's body.

If the person is standing upright or is walking on a treadmill, one or more support bars should be provided, especially if the VE may be on a slope, as the reaction of the person in the RE will tend to make them lean over in the opposite direction and, consequently, they may lose their balance and fall over.

There is a risk that the confusion between the virtual world and the real world could pose safety problems. Some people who are cognitively highly immersed in the VE may forget that they are in a simulated world. Inappropriate behaviours can occur. With our VR applications, we have frequently observed behaviour that is often more amusing than dangerous, as a user who is "too" highly immersed cognitively, sometimes places their control interface (joystick, etc.) on a virtual table that they perceive. And are then very surprised when it falls to the ground. On the other hand, a person who is very frightened by a virtual event may suddenly move backwards to escape the virtual danger and fall over. Attention must therefore be paid to the proposed content that can scare the immersed person too intensely.

9.8 CONCLUSIONS

This analysis of the disturbing sensorimotor discrepancies has allowed us to present in a clear framework a few solutions that, at the very least, will minimise some of the negative effects on the health and comfort of users. Some of these solutions will depend on the technical evolution of the VR headsets. Some have already been validated experimentally while others are simply avenues worth exploring. In the future we need to develop experiments to validate new solutions for the use of VR headsets by the general public.

We based our analysis on our $3I^2$ methodology supplemented by taking into account disturbances on a sensorimotor I^2 level to improve the health and comfort of users, which has resulted in its name: $3I^{2+}$. This analysis has its limitations as we examined each discrepancy independently and the assumptions made have yet to be justified and validated.

We did not explore the solutions for "communication" Virtual Behavioural Primitives to avoid making the presentation over-lengthy; however, specific solutions should be designed for communications between the VR system and the user wearing the VR headset. Compared to the use of a flat screen, wearing a VR headset places a significant constraint on users as they can no longer easily use a mouse and a keyboard.

The risks of the impact on the health and comfort of users of a VR application explain why manufacturers of VR headsets are issuing restrictive recommendations for the use of their products. And this for several reasons:

– they fear that the underlying problem of sensorimotor discrepancies in the virtual reality can be disruptive for the health and comfort of some users;

- the technical features of their VR headsets are not yet sufficiently developed and they fear negative opinions about their products;
- they are cautious and wish to protect themselves legally, particularly in terms of any potentially harmful usage associated with VR headsets in children.

Their main recommendations are: take breaks, stop when feeling unwell and, after using an application with a VR headset, refrain from carrying out any physically complex tasks such as driving. Children under the age of 13 are prohibited from using a VR headset. In the near future, it will be important to define precise rules of use applicable to VR headsets and analyse any **potential long-term effects** for all, especially for children, if the use of VR headsets becomes more widespread. The content of the VR applications should also be controlled and checked by using warning messages, as is the case with a certain number of VR headset manufacturers.

Even though manufacturers are cautious, they are also sometimes boastful and exaggerate the impact of their innovations: they are all able to best eliminate disruptive discrepancies. Such claims include the statement confirming that the use of the future Light Field screens will eliminate the "accommodation-vergence" discrepancy. However, this does not mean they should claim that they have solved the problem of simulation sickness, the "virtual reality sickness", as the other discrepancies have not been eliminated as yet.

Theories behind simulation sickness and motion sickness

The first theory is based on visual-vestibular sensorimotor discrepancies that are the cause of the sensation of feeling sick. This is currently the most widely accepted explanation for the appearance of symptoms caused by motion. This theory assumes that the orientation of man in a 3D space is based on a series of sensory stimuli from:

- vestibular systems, providing information on the movement of the head and its orientation in the terrestrial space;
- the visual system providing the body's orientation in relation to the visual scene;
- the proprioception of joints giving the relative positions of the limbs.

When the artificial environment (VR or transport) changes the coherency of all this information, such symptoms can appear (Harm, 2002). This theory is generally satisfactory but it does not explain certain specific situations such as the way users sometimes experience a sensory conflict without any sensation of sickness or how they are able to adapt to such conflicts.

Stroffregen and Riccio (1991) proposed a different explanation based on the fact that feeling sick as a result of motion in a virtual environment results from a discontinuation of the normal postural control mechanisms. The person receives conflicting stimuli that disrupt his way of controlling his postural position.

Treisman (1977) has used a theory on poisoning to explain why the sensation of feeling unwell as a result of sensorimotor discrepancies stems from an ancestral reaction to remove toxins from the stomach, in the same way as the brain perceives discrepant sensory stimuli when confronted with a situation of poisoning.

Part II

VR headset applications

Chapter 10

Introduction to applications utilising VR headsets

Before taking a detailed look at three specific applications utilising VR headsets, let us draw up a non-exhaustive list of sectors, both professional and non-professional, likely to be interested in VR headset use.

10.1 VR APPLICATIONS FOR ALL AGE GROUPS

- VR gaming and recreational activities. The critical question of how future VR headsets, marketed for the video game industry and related leisure activities, will affect the field is still up in the air;
- Digital arts (see Chapter 12);
- Social media. Interfaces could be developed to provide more human-centric exchanges;
- Telepresence technologies. VR headsets could be used for students, unable to attend school for health reasons, but who can be telepresent via a classroom robot transmitting images and sound, remotely. A greater sense of immersion, of being with fellow classmates in the building and on the playground, is created when classmates are seen via a VR headset (if the robot is equipped with a motorized camera, controlled by the movement of the absent student's head). But VR headset use for in-classroom teaching may be counter-productive and restrictive. While teaching in front of a large screen does not prevent teacher-student dialogue and allows all kinds of educational simulations, VR headsets should only be used in the classroom to meet specific educational needs, namely when a sense of physical immersion is useful (touring a campus or visiting a building);
- Drone piloting. Both VR headset features can be fully exploited but the impact of sensorimotor discrepancy is not to be underestimated;
- Marketing presentations. VR headsets can be used for product presentations that do not require aesthetic rendering. Certain complex end-user products, including the ergonomic layout of a future kitchen or that of the dashboard controls of a car, are a good example;
- VR video (360-degree video). An actual art form, and not just another film technique, utilising immersion in artificial worlds, both virtual (created with computer graphics) and real (filmed with 360-degree cameras). This new medium could be utilised for both private use, in line with family photos and personal videos, and public, though VR videos are experienced by the viewer, individually. Unlike film,

VR video is not director-controlled. Viewers participate in the VE and can affect the scenario; they determine their own point of view. It is not about telling a story, but "living" a story;

- VR music videos. Promising developments are being seen in 360-degree video with the advent of VR headsets with 3D spatialized sound capabilities, which were unsuited to simple headphone use;
- Sporting events. The 360-degree video transmission of sporting events is being envisaged, though classic television broadcasting is better suited to watching games, especially in the case of team sports; professional camera crews, and not the viewer wearing a VR headset, are better suited to choosing the best camera angles. VR video broadcasting could be used, however, to give the viewer a sense of the "pre-match or post-match atmosphere";
- Film viewing, during air travel or other means of transport. Lightweight VR headsets, not requiring head-tracking and with adequate image resolution, could be specifically adapted for travel.

10.2 PROFESSIONAL APPLICATIONS

- Social sciences. Virtual reality interfaces can be useful for carrying out behavioural experiments. By the same token, field scientists can contribute to improved VR headset utilisation by conducting research on user behavioural response in virtual environments. Such experiments are described in Chapter 10;
- Health: VR headsets, coupled with large and small screen VR techniques, are already in use in virtual therapy, in psychiatry, and in functional or sensorimotor rehabilitation. Certain surgical procedures, such as endoscopic operations, are better-suited to screen use. Augmented reality surgical interventions can be done via AR headsets, however.
- VR training and "serious games": Such techniques have been employed for 15 years and have proven highly effective without the use of VR headsets. No major changes in this area are to be expected with the introduction of VR headsets, therefore. These can nonetheless be tailored to meet specific visual immersion needs, keeping in mind the obvious drawback of visually separating the learner and the trainer (unless the trainer, too, is immersed in the same scene, requiring a more sophisticated RV application).

Transportation. Air, rail and ground simulators can be utilised via either large screens or VR headsets. Chapter 11 presents potential VR headset uses in industry.

Chapter 11

Behavioural lab experiments

Daniel R. Mestre

11.1 HOW VR HEADSETS CHANGE THE VR LANDSCAPE

If you consider the history of scientific discovery, the birth of immersive interactive systems is relatively recent. It is even more recent, if you focus on the use of VR as an experimental tool for the study of human behaviour (Mestre & Vercher, 2011). In behavioural sciences, researchers typically develop experimental devices, enabling them to control both sensorial inputs and behavioural outcomes in participants. In this general context, VR represented, in the 1990s, a significant step forward from two points of view, both linked to the development of computers' power. First, it enabled controlled, real-time interactive experiments. Secondly, it boosted the realism of experimental contexts. The general idea is that VR promoted the controlled investigation of "ecological" situations, and the validity of experimental observations, with reference to real situations (Loomis *et al.*, 1999; Tarr & Warren, 2002).

In the VR timeline, the invention of the CAVE (CAVE Automatic Virtual Environment (Cruz-Neira *et al.*, 1993)) constituted a significant step forward for the experimental implementation and assessment of ecological situations. Indeed, this multi-screen, human-scale, system enabled researchers to stimulate the entire human visual field and to provide participants with real-time feedback of their actions. Among the advantages of the CAVE that still persist today is the fact that 1) it enables real locomotion (within the physical limits of the setup), and 2) the participant naturally sees her real body (Figure 11.1).

However, awaiting the next technological breakthrough, this setup has numerous disadvantages. It needs a cluster of video projectors, which remains expensive to maintain and difficult as well to perfectly calibrate (in terms of geometry and colorimetry). It also needs (up to now) a cluster of high-end graphics computers, in order to deliver proper sensorial feedback to the participant's behaviour. Moreover, the latency of such systems (due to the path from motion tracking to sensorial rendering) is between 60 and 100 milliseconds, for a system whose basic working frequency is 60 Hz. This delay is usually not really disturbing, but can have a negative effect on fast gestures (participants often notice an instability of the virtual environment).

An alternative to the CAVE, which in fact existed before the CAVE, is the wealth of VR headsets that were produced since the "Sword of Damocles" (Sutherland, 1965). VR headsets have been around for a long time (just mentioning the early work at NASA in the 1980s). Until recently, VR headsets (at least those delivering a "decent" visual field) were quite heavy, with low spatial and temporal resolution. At that stage, for the

Figure 11.1 A participant in our 4-sided CAVE at CRVM (Mediterranean Virtual Reality Center[1]), offering a "walkable" space of around 100 square feet.

behavioural researcher, the CAVE remained the gold standard to conduct interactive experiments.

Something happened, in the last few years. Big companies put much money into small companies (Garage startups like Oculus). Today VR headsets are low-cost, compared to previous ones and even more so compared to video and PC-cluster based CAVEs. They certainly still have (for the most part) a limited visual field (around 100 degrees in the horizontal dimensions). However, one serious advantage they possess, as compared to a CAVE setup, is that their overall "perception-to-action" latency is in the order of 25 ms, thanks to a procedure called "time warping" (Evangelakos & Mara, 2016). VR headsets also have a specific "natural" property: They isolate the user's vision from that of the real environment. To achieve that using a CAVE, you need a 6-sided CAVE, completely surrounding the user. However, there are only very few of them in the world, given the technical constraints and prohibitive cost. For recent evolutions of the CAVE concept, see (deFanti *et al.*, 2010).

Recently, we started evaluating these new generation VR headsets in our lab, and we will here report a few observations that we made during recent experiments using them.

11.2 WALKING THOUGH VIRTUAL APERTURES

The scientific context concerns the assessment of behavioural presence in virtual environments (Mestre & Fuchs, 2006). Previously, we used our CAVE setup to try to

---

[1]www.crvm.eu

Figure 11.2 Virtual representation of the (avatar of the) subject passing though the aperture.

evaluate the behavioural aspects of presence in VR. The participant's task was to walk through a variable-width door (Figure 11.2). Results from this study (Lepecq *et al.*, 2009) suggest that participants spontaneously exhibited a behavioural strategy (rotating their shoulders to avoid collision with doorposts) that is close to what is observed during "real doors" crossing (Warren & Wang, 1987). We reproduced the identical experimental protocol when participants wore a VR headset (Oculus Rift DK2). In order to measure behavioural adjustments and to have precise corporeal real-time tracking, the CAVE tracking system (ArtTrack®) with eight cameras, using infrared recognition of passive markers, was used to monitor the subject's all-body posture (Figure 11.3).

As mentioned (see Figure 11.3), the participants were equipped with a body-tracking suit. This notably enabled us to study, besides measuring postural adjustments, the role of the vision of one's own body during behavioural interaction with a virtual environment. Indeed, when a user puts on the VR headset, he/she naturally loses sight of the real environment, including his/her own body. This is not the case inside a CAVE. We are facing a paradoxical situation. On the one hand, the headset is said to be more immersive (than a regular CAVE), precisely because it blocks the vision of the real environment, such that the user focuses his/her attention on the virtual environment. On the other hand, it creates a "disembodied" situation, in which the user loses sight of his/her own body. Given what we know about the role of the perception of our own body in perceptual calibration (e.g. Harris *et al.*, 2015), we suspected that the vision of one's own body might influence the participants' behaviour, and notably the perception of the environment's scale, thus the use of body-scale information in aperture crossing. Undocumented observations reveal that some subjects, when wearing a VR headset, have the feeling of floating over the ground, which might be due to the fact that they do not perceive the contact of their feet with the ground surface.

Figure 11.3 A participant wearing the VR headset, equipped with body markers, rotating his shoulder to pass through the aperture.

We thus used the headset device as a tool to study the role of a self-avatar in the "door-passing" task. As the user's body was tracked, we could compute and visualize, in real-time, a co-localized virtual body in the user's visual field inside the VR headset (Figure 11.4).

Initial results from this experiment are straightforward. First, the level of behavioural presence evaluated with the headset is similar to what was observed inside the CAVE. Subjects rotate their shoulder to walk through a narrow aperture (Lepecq *et al.*, 2009; Mestre, Louison & Ferlay, 2016). Secondly, when they cannot see (a representation of) their own body, they collide with narrow doors in almost half of the experimental trials. They appear to have problems calibrating the visual information to their own body size. In the presence of a virtual body, collisions are significantly reduced. The prime conclusion from this experiment, using a VR headset, is that behavioural presence, as indicated by adapted behaviour with respect to environmental constraints, requires a virtual representation of your own body.

Another observation from this study concerns cybersickness (Kennedy *et al.*, 1993). This symptom (commonly observed during VR exposure) appears to be due to sensorimotor discrepancies, inherent to VR technology (see chapter 8), linked to technical problems such as latency and/or tracking errors. It is observed in the CAVE, and the question was whether, as claimed elsewhere, the significant reduction of latency in today's headsets (as compared to the CAVE) could get rid of cybersickness.

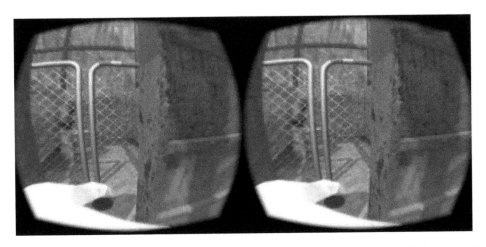

Figure 11.4 Subjective views (left and right screen) inside the headset, with the virtual body (in white)
The reader may try to free-fuse to see the 3D scene.

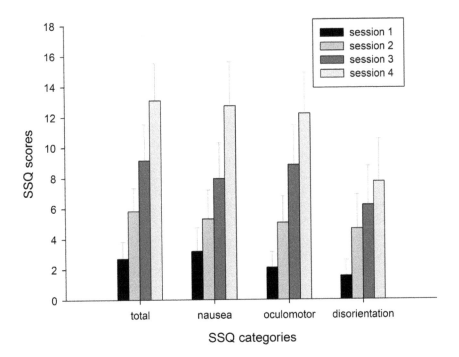

Figure 11.5 Average SSQ (simulator sickness Questionnaire) scores (for 18 subjects), for the four
categories and across the four consecutive experimental sessions.

During the experiment, participants were exposed to four successive testing sessions, interleaved with a resting period during which participants filled the Simulator Sickness Questionnaire –SSQ- (Kennedy *et al.*, 1993). Results are presented in Figure 11.5.

Results are clear and reveal a significant elevation of the SSQ score in all categories, along the four experimental sessions (each session lasting for about 10 minutes). There is certainly variability between participants (see the error bars in Figure 11.5). However, we can observe that all participants reported significant (superior to zero) symptoms from the second session onward (i.e. after 20 minutes of VR exposure). This result is coherent with the literature (Stanney et al., 2002). In other terms, the advanced technology of current headsets (at least the Oculus Rift DK2 we tested here) did not solve the VR sickness problem. Further analyses are however required, comparing all the different dimensions of differences between CAVEs and VR headsets (hard work) as well as investigating more precisely the different dimensions of the general concept of cybersickness (for instance, visual fatigue and eyestrain might not be correctly evaluated in the common SSQ).

11.3 CONCLUSION

Globally, our study confirms the capacity of VR headsets to induce presence in users. However, it also suggests that, as soon as spatial behaviour is involved (which is – almost – always the case), users need a virtual representation of their own body, in order to exhibit adapted behaviour with respect to virtual environmental objects (not colliding with narrow doorposts in our case). Moreover, this representation has to be spatially and temporally synchronized with the user's body, which is not necessarily easy, in technical terms.

Secondly, contrary to the common belief that latency was the main cause for cybersickness, our observations show that significantly reducing the overall system's latency (going from around 100 ms in a CAVE down to around 26 in a VR headset) did not abolish cybersickness. However, we can observe that SSQ scores seem to be lower than what was observed with a driving simulator (Mestre, 2016). More systematic comparisons are certainly necessary.

Finally, it remains to understand how multi-factorial determinants are hidden behind the general term of "cybersickness". For instance, it is now well recognized that VR headsets maximize the accommodation-vergence conflict: Stereoscopic stimuli trigger vergence eye movements while a fixed-distance screen blocks accommodation eye movements, both being naturally coupled. This might cause degradation of visual performance and/or visual fatigue. These factors seem to be insufficiently taken into account today and are certainly crucial for the development of "useable" VR headsets.

Chapter 12

Industrial use of VR headsets

Andras Kemeny

12.1 INTRODUCTION

While the first VR headsets appeared at the end of the 1960s, their industrial usage does not really begin before the 1990s. In fact the first VR headset, developed by Ivan Sutherland (Sutherland, 1968) used only wireframe images, insufficient for an acceptable immersive industrial experience. Real time computer generated images (CGI) with usable quality, i.e. with textured and anti-aliased images, only become accessible at the end of the 1980s with the Silicon Graphics workstations (Kemeny, 1987). PCs with nVidia graphics cards finally bring affordable real time image generation to the general public at the end of the 1990s, consequently leading to the development of industrial VR headsets, although their deployment remains confined to large industrial groups as their prices have remained high until recently.

The automotive industry expressed early interest in this new technology with Daimler comparing the VR based immersion quality with video projector display technology in real time (60 Hz) driver simulation tests as early as 1997 (Schiller, 1997), and then Volvo comparing VR and real scenarios by deploying a VR based driving simulator with nVision VR headsets in a Volvo 850 cockpit and measuring eye saccade durations (Burns and Saluäär, 1999). Both experiments support the relative validity of the systems used, but real production usage of VR doesn't really begin until a few years later at Renault, with a SEOS 120 dedicated VR headset integrated into a motion based driving simulator, which was specifically developed in the framework of the European Eureka CARDS project (Coates *et al.*, 2002). Once again, it is a VR based driving simulator, including a tracking system for better motion perception in driving conditions, which was deployed mainly for automotive human machine interface and human factors studies (Figure 12.1). This clearly announces the progressive convergence of driving simulation (DS) and virtual reality (VR), since confirmed by extensive data (Kemeny, 2014).

At Renault, following several projects for the integration into driving simulation systems, including the ProView VL50 VR headset with a 2.3 arc min resolution, to be compared with the 1 arc min human eye resolution, driving is judged difficult because of both the limited field of view (50° diagonal) and the CGI performance (Figure 12.2). Starting from 2004, with the evolution of the MOER project, first with SEOS, then Crescent, and finally with nVisor VR headsets, an augmented reality (AR) application was built and introduced VR and AR in the vehicle engineering design cycle

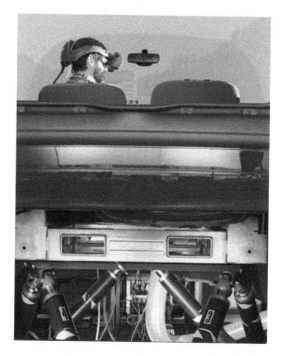

Figure 12.1 Dynamic driving Simulator with SEOS 120 VR headset.

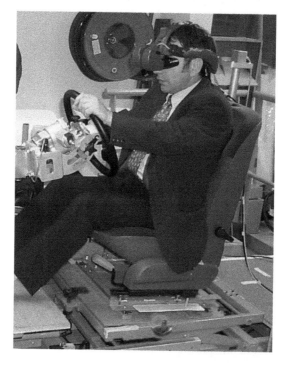

Figure 12.2 Driving using an ergonomic seat with a ProView XL50 VR headset.

(Kemeny, 2008). Hence the nVisor VR headset demonstrates an acceptable tradeoff between the LCOS technology provided colorimetric stability level and a sufficient large field of view, allowing to see simultaneously both windshield A-pillars of the project vehicle during the 360° driver vision evaluation. The instantaneous large field of view, offered only with a limited number of VR systems at that time, is a critical requirement, as it allows to simulate natural exploration of the surrounding environment without head movements which may induce time lags (transport delay) in the observation process (and presenting heavy user constraints with the cable equipped headsets).

12.2   DRIVING SIMULATION (DS) AND VIRTUAL REALITY (VR)

12.2.1   Convergence between driving simulation and virtual reality domains

The convergence between the driving simulation and virtual reality technologies dates from the very beginning of the advent of VR headsets, concomitant with that of driving simulators, both in the 1960s (Kemeny, 2014). Nevertheless, in a very short time these two domains come to use the same technologies, the computer generated images (CGI), and user interactivity with force feedback, producing user immersion both physical and mental, the typical characteristics of VR systems (Shermann and Craig, 2003).

The term Virtual Reality is introduced by Jaron Lanier in 1987 and the first immersive room installations (CAVE) appear in the following years (Cruz, 1992, 1993). The first industrial driving simulators deployed for car and traffic safety or engineering design are built around the same period (Allen, 2011; Drosdol and Panik, 1985; Nordmark, 1994). The CAVE systems arrive in France at the beginning of the 2000s, at Arts et Métiers ParisTech's Institut Image (2001), followed by PSA (Voillequin, 2006), and the first high performance driving simulators are built in the same years at BMW, Renault and Volvo (Kemeny, 2000, 2001).

12.2.2   Visuo-vestibular conflict

The driving simulation and virtual reality installations use similar technologies, consequently they encounter also similar difficulties in practice. The most common example is the simulation sickness or VRISE (Virtual Reality Induced Sickness Effects) when driving or navigating using interactive interfaces, such as a joystick or wiimote. This is caused in particular by the visuo-vestibular conflict, induced by the incoherencies between the human vestibular and visual systems during real time image generation without the production of the corresponding physical stimuli (Kennedy et al., 2001).

12.2.3   Transport delay or response lag

Another cause may be the transport delay or the lag in response between the original action and the corresponding rendered virtual stimuli, for example visual. This delay depends on the acquisition of movements notably that of head movements, image frequency (frame rate) and the motion platform for physical restitution, and may cause strong perceptual incoherencies (Berthoz, 2002).

While observing a scene whose images are rendered according to head movements, the rendering delays may induce malaise because of the temporal discrepancy between visual and vestibular cues. As the vestibulo-ocular reflex (VOR), which compensates head movements with eye movements and thus maintains the gaze stable on objects, is very fast, about 20 ms (Berthoz, 2002), there is compensation of head movement in generating images in the virtual environment (using VR headset, CAVE, cylindrical screens or other display systems) both for rotational movement and linear displacements. Combined movements are probably the worst (Angelaki, 2004), but their precise effect is not yet known completely.

Recent VR headset technology allows to deal efficiently with the lag generated by the visualization system, including 90 Hz (or better) display frequency if the images (CGI) are computed at a sufficient rate. All computing lags for acquisition and navigation devices are integrated into the global transport delay, which makes its effects critical for industrial use. For example, using a driving simulator equipped with a motion platform to reduce the visuo-vestibular conflict (see 2.2), the global transport delay is added to all additional computing lags for the diverse command control delay, in particular for steering and the motion system. These effects may be more critical for passengers in a driving simulator (Dagdelen et al., 2002).

12.2.4 Distance, speed and acceleration perception at scale 1

The absolute validity of simulation system's utility is not always required. For example, while studying human factors using driving simulation, the relative validity may suffice, meaning that while comparing different driving situations, the same differences will be observed as in real conditions (Kemeny, 2009). This argument can be made for human machine interaction applications with systems inducing the same mental workload as the simulated system, without the absolute validity in the rendering characteristics of the system. Nevertheless, speed and acceleration perception are crucial, as the mental workload and driver behaviour depend strongly on them (Barthou et al., 2010). Speed perception is a function of the visual field of view, because of the role of the human eye's peripheral vision, which is sensitive to movements and the level of visual immersion (Jamson, 2000).

Rendering physically accurate perception of acceleration when navigating or driving, in particular during rotational movements, strong braking or accelerating maneuvers, requires the use of motion platforms (Siegler et al., 2001). Usually only large dynamic driving simulators dispose of motion platforms, though high frequency and small actuator based systems (Mohellebi et al., 2004) may provide acceptable results, even with VR headset or CAVE virtual environments.

Perception of size and distance of objects in virtual environment is also crucial while using VR or AR systems, in particular for user evaluation of industrial systems for styling, ergonomics or lighting ambiance. Results in perception studies show differences between virtual and real conditions while evaluating distance and scaling (Loomis and Knapp, 2003; Kemeny et al., 2008).

One of the most important depth cues is the motion parallax, the role of which was shown sufficiently both in real and virtual environments (Rogers and Graham, 1979; Panerai et al., 2002; Kemeny and Panerai, 2003). However the precise role of the binocular vergence (Paille et al., 2005; Kemeny et al., 2008) and the cognitive factors (Glennerster *et al.*, 2008) are still being evaluated for the distance and scale perception in the virtual environment.

12.3 AUTOMOTIVE AND AEROSPACE VR APPLICATIONS

Vehicle architecture is one of the first automotive engineering design domains where virtual and augmented reality techniques were adopted. The objective is to evaluate architectural options for the driver station (or cockpit), in particular to check the impact of the windshield pillars on visibility. An immersive 3D representation of the vehicle architecture is necessary to correctly estimate the occultation of traffic hazards by the vehicle structure from the driver's point of view (Figure 12.3). 3D perception of the interior volumes and distances is also critical for cockpit roominess evaluation and perceived interior usable space evaluations, for example driver glove box accessibility and usefulness. However, visual rendering quality may be less critical for these domains than for styling design and perceived quality, for which very high display quality is required, at least 4K definition in most use-cases.

To replace the use of physical prototypes, which are costly and induce critical project development delays, architectural evaluation can thus be accomplished using corporate CAD data (Catia or NX), not only for interior visualization but also for ambiance lighting and headlight simulation, even from the early vehicle development phase (Kemeny *et al.*, 2008). The early industrial deployment of VR headsets occurred in France at Renault in 2004, as the systematic tool for architecture and ergonomics customer performance evaluation. A large field of view with more than 80° horizontal field of view (and at least 50% stereoscopic overlap) was required. Renault first

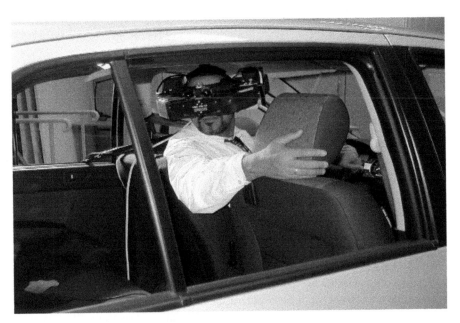

Figure 12.3 Vehicle interior vision with large field of view AR headset.

used a proprietary SEOS VR headset (120° × 40°) and later an nVisor system (SW111 – 102° × 64°), while continuing to test other competitor technologies, such as that of Sensics. This latter system is used by Airbus, though its multiple display unit calibration is comparatively lengthy and complex. Nevertheless Airbus implemented its maintenance VR application tool with the Sensics technology at the end of 2000s (Lorisson, 2010).

Augmented Reality (AR) headsets find a major application in vehicle architecture studies. With a composition of virtual environments with real and/or virtual driver cockpits, different engineering solutions were tested, for example using color key coding to distinguish vehicle interior and virtual environments displayed outside of the driver cockpit (cf. Figure 12.6). In addition, real vehicle images obtained by the cameras attached to the headsets (video see-through) can be easily compared to the digital representation of the project vehicle cockpit, thereby enabling diverse driver vision performance evaluations.

Nevertheless the advent of immersive rooms with 4K technology began the trend of CAVE VR replacing the VR headsets for industrial use. At the end of 2012, Renault inaugurated a state of the art 5-face high performance 4K display technology based on 120 Hz real time display (George et al., 2014), taking the lead from Jaguar, Ford, PSA and Volvo who are also currently using 4K technology based CAVEs (although some continue to test VR headsets for customer evaluation). Even Airbus, committed longtime to VR headsets, has now mostly switched to CAVE systems as its RHEA engineering design tool (for example for the A350 XWB design and development – see Airbus, 2012). High quality 4K projectors, based on TI technology (Christie) with high luminosity and high contrast performances, deployed with intra and inter collaborative engineering, makes these solutions cost efficient and attractive in spite of their high price. However, the advent of new VR and AR headset technology with very low image rendering lags even at high quality image rendering, such as the Valve of HTC or Hololens of Microsoft, announces another new evolution, undoubtedly bringing, in the foreseeable future, a shared and complementary use of the two types of technologies, to be presented in future reports.

The simulation sickness (cyber-sickness or motion sickness), experienced to some degree by all users but in particular female users (Flanagan et al., 2006), is still a major drawback for the use of VR in industrial applications, both with VR rooms and VR headsets (Colombet et al., 2016). In addition, VR headset intrusiveness is an important handicap, restricting both the utilization time, due to the general discomfort of the systems, and user movement, due to the heavy cables between the application PC and the VR headset (in virtually all models, except the Hololens).

12.4 SIMULATION SICKNESS (VRISE)

Simulation sickness is a critical factor in the use of VR systems as it limits the type of situations in which testing is suitable, and the duration of the design or validation experiment a user can complete. Several factors play a significant role. The transport delay is still a critical issue due to the amount of data to be displayed (i.e. several million surfaces in most automotive or aerospace applications). Nevertheless, a number of VR headsets are available on the market with a display lag of less than 20 ms with some applications, while using a display frame rate of 90 Hz or more (HTC Vive).

Other factors that play a critical role, such as user expectations, based frequently on the participation in previous experiments (Reason and Brand, 1975) or other physiological and psychological parameters that may influence postural stability (Riccio and Stoffregen, 1991). A past experience, for example navigating a narrow bend at high speed (with rotations), which resulted in a headache or nausea, may induce anxiety in a new experiment, even in conditions which do not represent other known stimuli for cyber-sickness.

From the numerous simulation sickness studies performed, some aim to define the acceptable physiological thresholds under which no undesirable effects are generated during the industrial virtual mock-up (VMU) evaluation, and provide data on the set-up conditions for industrial use-case studies. The results of several experiments carried out in Renault's high performance 170 M pixels 4K technology CAVE comparing simulation sickness effects in comparable CAVEs and VR headsets are now used for daily experimentation. The latest of these experiments compares two virtual reality systems: Renault P3I (Immersive and Intuitive Integration Platform) CAVE, a 4-sided virtual reality room powered by 4 ultra-short throw 1080p video projectors (Panasonic PT-DZ870) in a direct projection setup, with active stereoscopy (Volfoni active 3D glasses) and the Oculus Rift DK2 (resolution of 960×1080 per eye). While the Oculus rift offers a 110° field of view, the CAVE potentially offers a full field of view. However, the frame design of stereo glasses might partially mask peripheral vision. Simulation sickness was measured by SSQ questionnaires (Kennedy, 1993) and postural stability with a techno-concept Stabilotest (Riccio and Stoffregen).

The high acceleration levels (up to $13°/s^2$) induce significantly stronger SSQ values than the low ($2°/s^2$) levels and this both for expert and novice subjects (Figure 12.4). There are notably significant differences between expert and novice subjects, as the former almost never experience simulation sickness, occurring only rarely during strong accelerations, while the latter almost always experience simulation sickness during strong acceleration and may do so even with low acceleration levels (Kemeny et al., 2015).

It thus appears that novice subjects are frequently subjected to simulation sickness, especially while navigating and turning in tight bends, while experienced subjects almost never experience these effects, though it is well advised to keep the same low thresholds for experiments of longer durations for all subjects. The comparison

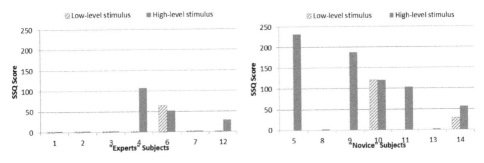

Figure 12.4 SSQ values for rotations: experts (left) and novices (right) subjects.

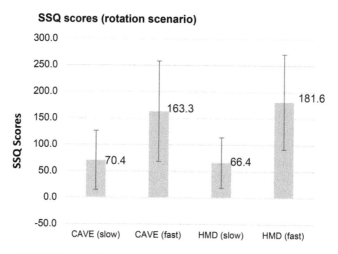

Figure 12.5 SSQ scores for both displays with the two rotational (yaw) levels.

between the CAVE and VR headsets (see above) does not show significant differences for the simulation sickness effects measured by the SSQ values (Figure 12.5) and (Colombet *et al.*, 2016).

Gender is an important factor for VRISE (see chapter 8) while training also plays an important role in avoiding motion sickness. Nevertheless, adaptation is progressive and its development seems to be linked to the specific situation experienced. As a consequence, new situations require new training (Reason, 1975). These results are to be supplemented by experiments with VR and AR headsets, though it would seem logical that the effects are comparable (Nelson *et al.*, 2000). However major differences may be observed between optical see-through and video see-through headsets, as composing images of different sources may induce perceptual differences, for example in distance or size perception (Kemeny *et al.*, 2008). Though optical see-through headsets have their drawbacks, the luminosity and colour dynamic of virtual images is significantly lower than that of real images. Hololens AR headsets are equipped with state of the art display systems, limiting these undesirable effects, even if Microsoft recommends only indoor usage. Dynamic calibration for a large range of illumination conditions is still an ongoing challenge for deploying AR systems in all lighting conditions.

12.5 SPACE AND SIZE PERCEPTION (SCALE 1 PERCEPTION)

One of the major challenges in the industrial use of VR systems is the correct perception of scale (Kemeny and Panerain, 2003). Loomis and Knapp (2003) report perceptual differences when comparing virtual and real environments. A more recent study confirms these results, though with smaller differences, using a stereoscopic display with different field of view and observation conditions (Paille *et al.*, 2005). According to these studies, distance perception is undervalued, meaning objects in virtual environment are perceived to be larger than in real conditions and furthermore the smaller

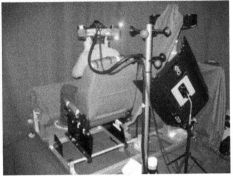

Figure 12.6 Observer in face of a cylindrical screen and wearing a VR or AR headset.

the field of view, the larger the misrepresentation will be. These results are coherent with those observed in the real environment with different fields of view (Alfano and Michel, 1990; Hagen *et al.*, 1978; Watt *et al.*, 2000), though the issue stays open and in intense discussion concerning the virtual environment (Knapp and Loomis, 2004; Stefanucci *et al.*, 2015). It appears that the use of motion parallax and its quality strongly influences depth perception (Gogel and Tietz, 1973; Howard and Rogers, 1995; Kemeny and Panerai, 2002).

To compare the impact of the type of display system, scale perception was evaluated with different VR systems at Renault in the late 2000s (Kemeny *et al.*, 2008). One of the studies produced results on scale perception using three different display systems:

- SEOS 120 VR headset with a 120° × 40° field of view (50% stereoscopy) equipped with Sony (768 × 582) cameras aligned with observer eyes used in AR mode (video see-through, see Figure 12.6),
- The same VR headset used in VR mode (no vision of the human body) with (VR HT) and without (VR no HT) head tracking (see Figure 12.6),
- A cylindrical screen (Galaxy Warp Barco projectors with 210° × 50°, see Figure 12.6).

The observer in this case compared the perceived size of Renault Scenic cockpits, one being the real object and the other a representation in the virtual environment, and these in a random order to avoid undesirable learning effects (see Figure 12.7).

Results show that scale perception is comparable between real and virtual environments with VR headsets, while size is overestimated using the cylindrical screen without head tracking (that is distances are undervalued), and size is underestimated in augmented reality (Figure 12.8).

The lack of head tracking well explains the overestimation with the cylindrical screen (Rogers and Graham, 1975), as head tracking is only possible using real time distortion correction algorithms with cylindrical screens (Szymanski, 2002), available at Renault only since 2008 (Filliard *et al.*, 2010).

Figure 12.7 Virtual Cockpit: scale I, inferior of scale I, superior of scale I.

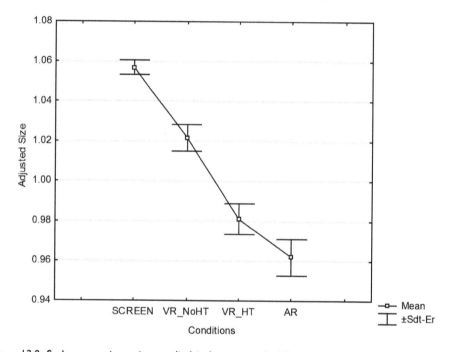

Figure 12.8 Scale perception using a cylindrical screen and a VR headset with and without tracking in immersive or augmented reality mode.

The strong underestimation using the AR headset which allows perception of the human body seems somewhat counterintuitive, as the presence of the human body provides visual scale information (Rock and Harris, 1967) and influences size perception while observing the surrounding environment. However, the results of previous studies show that the overall luminosity in the observed real world environment may influence distance perception (Coello and Grealy, 1997), which may be the cause of discrepancy due to the different image sources in the case of video see-through headsets. Another explanation of the impaired perception may be the result of the differences in providing stereoscopic real world images captured by the camera and virtual world images computed by the image rendering system (Hale and Stanney, 2006).

Optical see-through headsets may not have the same source of the impaired scale perception observed with the video see-through headsets, nevertheless the luminosity differences between the real world and the displayed images may still induce notable perceptual discrepancies. In addition, while representing complex, rich environments, the differences in stereoscopic viewing (different vergence for different real world objects and a unique vergence plane for virtual objects) remains a major challenge for the current AR headsets. The conflict between vergence and accommodation, due to the binocular and accommodation distance, also induces fatigue and simulation sickness effects (Hoffman *et al.*, 2008; Watt *et al.*, 2005).

In conclusion, the integration of all the visual cues and cognitive factors is a domain of intense research, promising new results and a better understanding of visual perception in the use of VR or AR headsets.

Chapter 13

Creating digital art installations with VR headsets

Judith Guez and Jean-François Jégo

Virtual Reality allows us to get immersed in artificial worlds, and to interact in real-time in these universes. The potential of this technology is so vast, as it lets create either realistic, symbolic or imaginary worlds, that it is not surprising to notice that artists have been using it for several decades. Virtual Reality as an artistic medium allows the spectator-user to live immersive and interactive experiences, and it gives the opportunity to the artists to explore new ways of expression. In this chapter, we first describe the added value of the VR headsets in artistic creation, in regard to a historical approach and other interfaces. We then propose a creation model to produce artistic installations making use of the VR headsets' characteristics.

13.1 VR HEADSETS IN ARTISTIC CREATION

13.1.1 Virtual Reality as an artistic medium of creation

13.1.1.1 The relationship between the spectator and the artwork in art history

Immersion and interaction—the two inherent notions in Virtual Reality (Fuchs *et al.*, 2006)—were questioned and experimented by the artists before the arrival of the digital era, playing with human perceptions and senses, and allowing the spectator to interact with the artworks. In this part we propose a brief overview of the use of these concepts in modern and post-modern art movements.

In the late 1950s, the environmental art starts to bring art out of the museums, meeting the spectator in his living space, surrounding him and being a part of his environment. It proposes to include the spectator in an artistic space which can be installations[1], or modified natural spaces or landscapes[2] with the intention to place the spectator at the center[3].

In the same period appears the "participative" artistic movement where artists offer further involvement of the spectator in the artwork, directly taking into account

---

[1] *The Cyclop*, 1969–1994 by Jean Tinguely and Niki de Saint Phalle.
[2] *Land Art* artists: Mike Heizer, Dennis Oppenheim, Dan Graham.
[3] The *Environmental Art* proposes to isolate or surround the spectator "the arrangement of an interior space, where the art object can enter in relationship with the environment" (Popper, 1985).

its relationship with the work. Art becomes more accessible, getting closer to the public (Popper, 1985). This relationship can be considered as a full-art form with its own rules (Tramus, 2001). It has more recently been conceptualized as *Relational Art* or *Relational Aesthetics* (Bourriaud, 1998). For instance, artists from Optical Art and Kinetic Art Movements propose to explore more deeply how the spectator may be attracted to the work. They question the instability of perceptual mechanisms (Couchot, 1998), in order to get closer to the public. For instance, the work of Marcel Duchamp, *Rotary Glass Plates* (Duchamp, 1920), invites the spectator to activate the glass plates turned by a motor and he can discover a moiré pattern given by the rotation. This is establishing a new relationship between perception of the artwork and the possible action on it.

13.1.1.2 *Virtual Reality in digital art*

New ways to immerse the spectator in an artificial environment come up with the analogic era, especially in *Video Art Movement*. Then appears in the 1970s *Interactive Art Movement* in which the spectator can interact with artworks which can react in real-time through first analogue and then digital interfaces. For instance, Myron Kruger in the analogic installation *Videoplace* (Kruger, 1970) allows the user to draw with his own silhouette. Interaction is becoming more and more in the field of art an opportunity to artistic experimentation.

In the 1980s artistic installation appears and proposes that the spectator can "control" artwork using buttons for instance[4]. This recalls the approach of Marcel Duchamp in the work *Rotary Glass Plates* (Duchamp, 1920). Then, in the 1990s, the interaction becomes more relational, proposing responsive artworks in real-time and in which the spectator can become an actor of the artistic experience. In the meantime, we observe the emergence of Virtual Reality devices such as VR headsets and CAVE systems (Cruz-Neira, 1992). In 1995, the artwork *Osmose* (Davies, 1995) created by Charlotte Davies allows spectators to explore real-time artificial landscapes inspired by the nature and by the life, using a breathing sensor built-in a belt and a VR headset. The installation *World Skin* (Benayoun, 1997) by Maurice Benayoun uses a CAVE system where the spectator can play the role of a war photographer. Taking a photo using the "camera" interface that is tracked in real-time is literally tearing off pieces of the virtual world and replacing with white holes. After the experience, the spectator can retrieve a printed version of the picture taken. The artist questions the medium itself by its hybridization and its connection to the reality.

In the 2000s, several artists inspired by the life create new relations between the spectator and the artworks, leaving room for improvisation and "a living and freer art" is emerging (Guez, 2013). In the artistic installation *The Virtual Tightrope* (Bret *et al.*, 2000–2006), the virtual funambulist is not only responding to stimuli of the spectator to restore its balance, but it takes into account its environment and adjusts its posture in real-time according to his perception, using machine learning with neural networks. The authors of the installation, Michel Bret, Marie-Hélène Tramus and the neurophysiologist Alain Berthoz proposed a new paradigm called "second interactivity paradigm" (Bret *et al.*, 2005) in reference to the second cybernetics where interaction is no more

---

[4] See the artwork of Jean-Louis Boissier *Pekin pour mémoire* (Boissier, 1986).

Figure 13.1 Example of the artistic installation InterACTE; a) improvisation with the shadow of the virtual Pierrot, b) immersive improvisation in the 3D virtual world using the VR headset, c) view from the VR headset.

external control, but emerges from a relationship between the user and the installation. This new kind of relationship can be an opportunity to create new artistic forms.

Virtual Reality turns out to be a wonderful medium that allows thinking and giving shape to different forms of immersion and interaction. Artists can play with its potentials and its limitations to better customize the medium and thus enhancing it[5]. Here lies the challenge of creation!

13.1.2 Characteristics of VR headsets in regards to other interfaces for artistic creation

Virtual Reality offers a lot of sensorimotor interfaces (see Chapters 2 and 6) and each year brings innovation which is interesting to explore. For the last decade, many VR headsets reached the market, cheaper and therefore more usable and accessible. They each have different characteristics, quality and technical limitations that have to be known. In fact, besides the influences on sensorimotor and cognitive aspects, they offer creative potentials for many experiences different from the ones made possible using screens, video projections or a CAVE system.

13.1.2.1 Isolating the spectator from the real world

An important property of VR headsets is its ability to isolate visually and acoustically the spectator from the real world. The subject can be completely immersed in 360 degrees in the virtual environment. The position and orientation trackers (of the head, the hands, the feet … only or even the whole body) allow the user to navigate and to interact with the virtual environment around him. Stereoscopy allows depth effects that should be well configured and dosed to avoid discomfort (see Chapter 9).

This isolation allows the spectator to feel more "free" to experiment *in-camera*, driving his own observation, without feeling the other spectators have all eyes on him. In the installation *InterACTE* (Batras *et al.*, 2015) the spectator is invited to interact with a virtual Pierrot during two acts. In the first act, he can improvise with the shadow of the virtual character performing gestures with his arm (Figure 13.1a). In the second

---

[5]To learn more about Digital Art, see The DICCAN (the dictionary of digital art) by Pierre Berger which provides a complete list of artworks (Berger, 2010).

Figure 13.2 Example of the artistic installation Lab'Surd: the LABoratory of SURvirtuality; a) scenography and set in the real world; b) VR headset point of view at the beginning of the experience: the virtual set is a copy of the real set; c) VR headset point of view at the end of the experience: the set becomes abstract.

act, the spectator is immersed in the virtual world of the Pierrot using a VR headset (Figures 13.1b and 13.1c). The authors noticed the spectators tend to more "let it go" when they improvise immersed in the VR headset than while they were improvising with the shadow. Indeed, in this case they could perceive the public observing (Batras *et al.*, 2016).

In the VR headset installation *Lab'Surd: the LABoratory of SURvirtuality* (Guez *et al.*, 2015), the spectator is immersed in a virtual room that is a copy of the actual room where he is sat down (Figure 13.2). While the spectator wears the VR headset, he gets immersed into another world he is free to explore (a). This virtual world is at the beginning of the experience a realistic copy of the actual room where the installation takes place (b). Then, the place will progressively be transformed and deconstructed until becoming more abstract with paradoxical architecture (c). The spatial sound aims to guide the spectator and make him look around him. The sound is revealed to be an important element to use to guide the eyes and the narration.

13.1.2.2 The spectator in an invisible body

We note there is a fundamental difference between a VR headset and a video projection system or a CAVE: the user does not see his own body. The subject immersed will naturally use his other senses such as touch and proprioception to perceive the environment or his body using palpation or feeling around (see Chapter 3 for a detailed description). From an artistic point a view, it may be interesting to play between what the user can see or feel focusing on the senses of vision and proprioception. However, the creator should always pay attention to sensorimotor conflict (especially between the vision and the vestibular systems) that can be the source of discomfort or cyber sickness (see Chapter 8).

In the interactive installation *Lab'Surd* we have presented above, the spectator is sitting in a chair and puts on the VR headset. In the virtual world, he can't see his own body. The authors observed that this feeling could be experienced as strange by the spectator at the beginning of the experience. The spectator grabs a real plastic glass whose virtual copy follows the movements of his hands. This object allows to interact with the virtual world and also to feel the real set. The glass becomes prosthetic as a visible extension of his invisible body in this new artificial world. Furthermore, the authors observe that as soon as the user grasps the glass, he finds his bearings and accepts the absence of his body in the virtual. Thanks to this trick, things that we could consider like incoherencies (not to see your body) becomes only ambiguities between the real and the virtual, and they now participate in reinforcing the surreal and the mystical aspects.

13.1.2.3 Accepting a virtual body

It is a fact we can't see our own body in a VR headset, but it is possible to overcome this. A simple solution consists in placing the spectator in a virtual body which could be a scan or a model of his original body or a different one as a new skin or a suit. This virtual body would be better accepted if its movements are following the spectator's movements using for instance a motion-capture system. Several studies showed the virtual body does not have to be an accurate copy of the real one since it appears that the spectator will quickly "accept" and adopt this new envelope or avatar (Slater *et al.*, 2009).

In the art installation *Liber* (Guez *et al.*, 2015) the spectator immersed in the virtual world is in the shoes of a human-plant hybrid character placed in front of a virtual mirror. The arms and head of the spectator are tracked in real-time allowing him to solve riddles helping with the tattoos placed on his virtual body. The spectator has to refer to his vision, his proprioception and has to touch his own body to interact with the tattoos. Playing with the ambiguities such as presence-absence or likeness-difference from his own body, allows for many varieties and seems a very interesting concept to explore for creation.

13.1.2.4 Designing interaction with the VR headset

The VR headsets are usually equipped with a tracking system of the user's head which can be used for interaction only, allowing not to overload the user interface.

Navigation in a virtual world without moving in the real world

The orientation tracking of the head can be a very simple way to navigate in a virtual environment, especially if the virtual camera follows a path. For example, in the work *I need a Haircut* (Fleuryfontaine *et al.*, 2014), the camera is looping on a rail in the virtual world. Since the visit is looping, the spectator can put on the VR headset and dive into the artist's world at any time of the experience. He can remove as well the helmet and leave the virtual environment whenever he wants. The authors use here a very simple way to navigate, and the true interactive action here is the observation. This approach is very similar to many immersive installation focusing on interactive observation like 360 interactive video.

In the music rhythm Virtual Reality experience *Playhead* (Auxietre, 2015a) designed for the Samsung GearVR headset, all the interactions in the virtual world are based on the orientation of the head, using a sight at the center of the image that allows several types of actions (pressing buttons, modulating music tracks, etc.).

In the same way, the author of the installation *The 5th Sleep* (Auxietre, 2015b) proposes an artistic exploration of the human body. The spectator can observe around him and drive the virtual capsule in which he is seated by turning his head and pointing at the path he/she wants.

Moving in the virtual world the same way we physically move into the real world

Some installations using a VR headset have a tracking area that allows to move on a large surface (from 2 to 9 square meters). The spectator is able to move freely and walk in a virtual space the same way he walks in the real environment. In this case, there are many fewer visual-vestibular inconsistencies (S21 solution, see chapter 9).

Control and action

Some VR headsets such as the Samsung GearVR have buttons that can be used to perform actions in the virtual world (like pressing virtual buttons, manipulating virtual objects, navigating, etc.) or application control (resetting or exiting an application without removing the VR headset). We note that the microphone can also be used to trigger events. In the Virtual Reality game *Panopticon* (Spiers *et al.*, 2015), the user has to whistle to control the prisoners of the prison. Of course the microphone can be used as usual to communicate with virtual agents or with other persons.

Beyond navigation, the tracking system of the VR headset gives a good estimation of what the spectator is observing and this is useful to trigger events depending on what he is looking at (or not). *The Sightline experience: The Chair* (Mariancik, 2013) for Oculus Rift uses this technique to change the objects and the places when the user does not look at them. This method was also used in the installation *Lab'Surd* to move objects around when the spectator does not look at them, playing with visual and surprise effects.

We note some VR headsets start to be equipped with eye-tracking systems (see chapter 7). This will allow to detect the gaze and better know what the user is observing, and this gives interesting perspectives of interaction for creation.

13.2 A METHOD TO CREATE WITH A VR HEADSET

13.2.1 The Diving Model

The choice of a VR headset to create an interactive artwork must take into account the goals of the experience, and the narration proposed by the artist. Each artist can build an experience that corresponds to his world. The artist can use the two specific features of the VR headset (visual immersion and head tracking) to better imagine how to build the scenario playing with time and space. Basing our observation and analysis on previous artworks using VR headsets, we note recurring patterns.

First, the spectator will live the artwork by himself. The experience should be designed for his point of view. The presence effect felt during the experience in a virtual or a mixed world can thus be conceptualized "as an artistic form of artwork-spectator relationship," and thus take into account several creative issues (Guez, 2015), like:

1. How to engage the spectator in his first contact with the artwork?
2. How to maintain this bonding?
3. How does he leave the experience? And what effects or memories will he keep?

To answer these questions, we propose a three-steps design: the immersion protocol (a), the presence in the virtual world during the experience (b), the emersion protocol

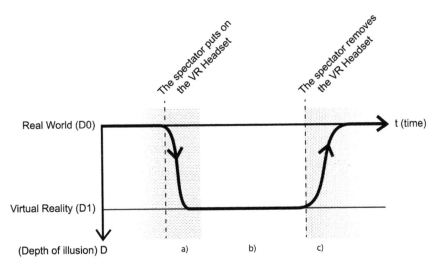

Figure 13.3 Diving Model: diving into a virtual environment using a VR headset, translated from (Guez, 2015); a) immersion protocol; b) experience: presence in the virtual world; c) emersion protocol.

(c) (Figure 13.3). These steps can be depicted using the "diving model into the virtual world" (Guez, 2015), as a guide for artistic creation. The following illustration shows how an experience can be designed following a timeline (t), from the spectator's point of view. It identifies the moments of transition between the experiences of the spectator in his usual "primary" reality (D0) and his experience in Virtual Reality (D1). Here, the VR headset allows to be the passage between the two worlds: to allow diving into the depths (D) of another realistic world, imaginary or symbolic.

This schema is proposed as an empirical support to design an experience using the VR headset and allows each spectator to interpret, to perform or to live his own experience. It is interesting to collect feedback from the spectator to compare the impact of the artwork with the artist expectations. These observations are a true source of inspiration to continue to explore Virtual Reality as an artistic medium. In the next part, we describe in detail these three steps: the immersion protocol (a), the presence during the experience (b) and the emersion protocol (c).

13.2.2 The Immersion Protocol: Scenography and Storytelling

The first contact with the artwork is important: the spectator is invited to leave his usual reality to the "reality of the artwork." The work has to make credible the context of the experience and to invite the person to accept to dive in, according to the principle of the suspension of disbelief proposed by Coleridge (1817).

13.2.2.1 Create the real set of the artwork

It seems essential to design the space in which the artwork is. The author has to think how and where the spectator will be placed, how he is coming in, how he is moving (physically), and finally how he is leaving the experience. For example, if the storytelling of the installation takes place only in a virtual world, a simpler scenography can be designed, using for instance a single chair in a quiet space. However, if the storytelling of the installation is hybridizing the real space with the virtual world, it is important to set up the VR headset and the interfaces according to the design of the scenography. We propose three different types of position of the spectator in the artwork:

- A lying position invites to stay more static so it is more adapted for a contemplative experience;
- A sitting position doesn't allow much movement, using mainly the upper body (arms, head), manipulating, exploring, etc.;
- A standing position allows to move around the real space and can engage more to manipulate, explore, walk, jump, search around, etc.

According to different types of scenography, we propose two example situations:

- The person is standing, sitting down or lying and puts on a VR headset visible in the place of the exhibition. He is less isolated from the real world and the other spectators can see him;
- The person is standing, sitting down or lying and puts on a VR headset located in an isolated space in the exhibition (with curtains or in a dedicated room, etc.). Spectators can enter only one by one.

13.2.2.2 Taking into account the exhibition space

The exhibition of the work has to be carefully taken into account. If the installation is made for a specific space, it will fit the architecture of the place. However, if it is shown in another place, it may have to be adapted.

When the same installation is exhibited in different places, it is important to know that VR headsets have their technological constraints which may interfere with the artistic purpose. The scenography has to be designed taking into account for instance the cable length, the placement of tracking cameras or the other interfaces, the presence of direct light regarding infrared systems, etc.

Also, in the case there are other VR headset artworks or demos in a shared space, it seems important to check the technical issues or interface and to require as much space as possible between the installations in order that the spectator can experience in the best conditions.

The presence or absence of a mediator raises important questions: does the installation require a mediator to help for instance the user to put on the VR headset? Is the installation working autonomously? In all cases, it seems important to propose a manual or technical descriptions especially if the author is not present during the exhibition.

13.2.2.3 Creating a scenography which helps the transition from the real to the virtual world

We have previously shown different types of scenography to design the exhibition space. We propose here to use this space to create a transition from the real to the virtual world. This allows the spectator to better dive into the virtual reality experience (see diving model, Figure 13.3). The place has to attract the spectator, to question him or even to encourage him to start to live the experience before putting on the VR headset. This place may be:

- completely real;
- real and virtual, for example with video projection mapping on real objects;
- mostly virtual, for instance the spectator wears a VR headset in the middle of a video projection (projection wall see Figure 13.1 InterACTE or CAVE systems).

The transition from the real to the virtual world, which happens when putting on the VR headset, can hybridize properly the real physical space and the virtual environment. The moment the user puts on the VR headset can also be used for artistic purpose, creating for instance:

- a progressive transition designing similar real and virtual environment;
- a progressive transition showing a virtual body that replaces the actual body of the user;
- a rapid transition without any correspondence between the two environments.

13.2.3 Presence during the experience: maintaining the relationship

Putting on the VR headset brings the spectator at the heart of the experience. We will focus now on maintaining consistency regarding the storytelling using the presence effect we will describe in the next part.

13.2.3.1 The different styles of storytelling

We propose three different types of scenario for a better immersion of the spectator in different worlds:

- following a linear narrative pattern;
- creating a story with multiple choices;
- creating a free environment where the spectator can choose to interact and/or where he can take the time to contemplate or to explore.

The style of the experience can adopt a figurative or a totally abstract form, leaving the spectator to choose to explore, to wander around, to contemplate or to improvise, etc.

13.2.3.2 Keep the spectator present in the virtual world

Jaron Lanier defines Virtual Reality as "a computer-supported way of creating the illusion of being in an alternative world with other people. It's a sort of dreaming you do consciously that other people can be a part of" (Lanier, 1990). The concepts of "illusion" and "being in another world" bring us to a specific concept called Presence in virtual reality. Many studies focus on the presence in virtual reality to define it more precisely (Slater, 2003; Nannipieri et al., 2014; Bouvier, 2009). For instance, psychologists used it successfully in the design of cyber therapy experiments and the analysis of their effects (Wiederhold and Bouchard, 2014). Guez (2015) proposes to use the presence as a tool in the creation of artistic experiences. It is commonly advisable to keep the presence at a high level during the experience[6]. But this remains difficult due to technical or cognitive reasons; research is still in progress. Also, it is not usually recommended to break the presence while the spectator is immersed, to do so, we recommend for instance not to distract him in the real world, designing a user-friendly interface. Although, we can imagine the break in presence can be part of a desired artistic effect.

13.2.3.3 Playing with the concept of presence

We previously saw it is not recommended to break the presence effect since it will cause the spectator to pay more attention to the real world. However, it may be interesting to play with the concept of presence itself. The work of Mel Slater (Kilteni *et al.*, 2012) provides many examples of the possibilities of illusions in the virtual world. From an artistic point of view, it could be useful to create intentional breaks making the spectator pay simultaneous attention to the real world and the virtual world. For example, in the installation *Lab'Surd*, the spectator goes from virtual space to virtual space as a *mise en abyme*. Here, the challenge lies in creating coherent transitions, but also with deliberate "breaks in presence" when he moves from one environment to another. The purpose here lies in better making him conscious of its current experience. The model presented in Figure 13.3 that was used as reference can also be used to go further in *mise en abyme* plays[7].

13.2.4 The Emersion Protocol

The end of the experience has to be carefully designed since it can be both an important part of the artistic purpose and have an impact on the user. Some artists are even using the concept of emersion as an artistic purpose (Sohier, 2016).

13.2.4.1 The last scene: transition from the virtual to the real world

In contrast with the immersion protocol that creates the passage from the real to the virtual world, the emersion protocol allows the transition from the virtual to the real world. It creates a link between the last scene experienced in the virtual world and the return to reality. Depending on the artistic purpose, several types of emersion are possible. We can use for instance a fade out or a black screen like a movie ending. It can also not be different from the experience itself and the spectator can leave whenever he wants. The return may thus be more or less progressive or abrupt.

It is important to note that this return can confuse the spectator. Disorientation can lead to side effects due to sensorimotor inconsistencies in the virtual environment. The emersion protocol can help prevent this by helping the spectator to progressively

---

[6]To learn more about presence in artistic video games and VR, see the interview of Jenova Chen (Huon, 2015).
[7]See Guez (2015): different examples of creations and diving models related to the concept of oscillation between the real and the virtual worlds basing on the depth of illusions.

return to reality, using for example fade out of the sound or the image or by allowing to take the time to get up.

The scenography can also help the return to reality the same way it invited the spectator to enter the experience. We propose some smooth transitions to the real world such as a different sound or ambient light from those at the beginning of the experience. Finally, the author, the mediator or the actor in the case of a performance can assist "awakening" to return to reality.

13.2.4.2 Back to reality and collecting feedback

That's it! The spectator removed the VR headset. He leaves the installation space but it doesn't just end here. Just out of the installation, the memory of his experience is still there. Depending on how he lived the virtual journey, the spectator could want to tell what he experienced and this could contain precious information for the artists. It could be rewarding for the artists in order to continue the exploration of the medium and the artistic process. The qualitative method of the "explicitation interview" (Vermersch, 2012) is a recent but efficient method to deeply understand a specific moment we experienced[8]. Here this method can be used to help the spectator explain his memories and thus raise meaningful and sensitive information to better create new experiences (Batras *et al.*, 2016). This could be a way to get closer to the essence of the aesthetic relationship between the spectator and the artwork.

13.3 CONCLUSION AND FUTURE DIRECTIONS

Virtual Reality is a medium that offers stimulating opportunities for artists. The VR headset allows the author to bring the spectator into fictional or reinvented worlds, in which space and time can be completely different from our reality: it allows to experience another reality, and thus to question our own perceptions.

In this chapter, we mainly presented examples of artistic installations using virtual reality. However, the VR headsets can be used in performing art, such as theatre and dance. For example, in the theatre play *Cassandra-Matériaux* (Chabalier *et al.*, 2016), the actress is playing the role of the Greek prophetess Cassandra, who is assailed by visions of the future expressed by the god Apollo only she is able to see (Figure 13.4). The VR headset on stage is the gateway to the world of visions of Cassandra: when she wears it, she is able to see the virtual Apollo, and can express her visions. During this scene, the actress appears to be the observer of a virtual world, and the spectator can imagine what she is seeing only through her, creating a *mise en abyme* on stage. Several scenic games can then be imagined, playing with what is visible and what is not.

In the field of cinema, a few film directors are also experimenting with this medium, particularly since the development of 360 degree videos. For example, the director Chris Milk has been making movies for VR headsets since 2013 and regards this new medium as an empathy[9] machine (Milk, 2015). In the same way, the artistic experiment *The Machine to be another* (BeAnotherLab, 2015) proposes to swap our body with

---

[8] See Guez (2015): Method for analyzing audience feedback with the explicit Interview record.
[9] To learn more about the concept of empathy in digital art, see Bret *et al.* (2005) and Batras *et al.* (2016).

Figure 13.4 Theatre play Cassandre-Matériaux. The actress (in the centre) is immersed in the virtual world using a VR headset, the audience can perceive here and there what she is able to see in the VR headset with the video projection on a tulle (on the right). Photo by Emmanuel Valette.

the one of someone else, by showing a video stream of the other person's point of view. Live video feed makes the creation of new experiences possible. Also, the real images of the video stream can be hybridized with virtual elements (see the field of augmented reality and augmented virtuality[10]).

Virtual Reality makes it possible for the artists to create new experiences and explore new uses of VR headsets in many different ways. The device can be considered like an (aesthetic) object, questioning the possible ways to get immersed or to interact, and keeping hybridizing the medium with tangible and intangible materials. All of this constitutes a wonderful creation material for the artist so that he could freely define his *Art*.

ACKNOWLEDGMENT

The authors wish to thank Philippe Fuchs for the initiative of this book, Marie-Hélène Tramus and all the people from the INREV laboratory at the University Paris 8 who contributed to feeding our reflections. We also would like to thank the proofreaders Pierre Berger, Guillaume Bertinet and Antoine Huon. This chapter was written with the support of the VRAC: Virtual Reality Art Collective (http://www.vracollective.com).

---

[10]See Jin-Yao Lin *Sensation World*: the artistic installation using a VR headset to see the live video stream of a fish in an aquarium mixed with 3D printed objects (Lin, 2016).

Chapter 14

Conclusion and perspectives

14.1 CONCLUSION

In the opening chapter of this book, a certain number of issues was addressed regarding the behavioural response of users wearing VR headsets, and in particular the effects of sensorimotor discrepancy on user behaviour. Potential solutions to some of these issues were also explored, as were questions regarding the likely introduction of VR headsets to the general public. Our aim, however, was not to provide a complete review of such open-ended questions. Certain issues, such as the eventual links between different kinds of sensorimotor discrepancies, were not addressed.

The fact is, VR researchers today do not enjoy the benefit of hindsight, the techniques of the sector having only emerged in the past twenty years, and they have an even greater lack of perspective in the case of VR headset use. Consequently, future applications designed for the general public will inevitably be based on partial scientific data. Developers will have to base their work on user trials, but such experiments will only concern a small number of participants whereas a sizeable tester population is needed to establish conclusive findings. Furthermore, a host of other parameters will need to be tested, and scientifically controlled, in order to determine valid solutions for mass-produced applications. In other words, behavioural response assessment will require costly time-consuming studies, to which developers will be unable to devote themselves. Additionally, a large number of tests will have to be conducted for accurate and irrefutable conclusions to be drawn.

It is therefore up to researchers in virtual reality and, in particular, the field of cognitive sciences, to continue carrying out experiments focused on the particularities of this new visual interface and its future consumers. Cognitive scientists, especially those in the field of neurosciences, need to answer questions of behavioural response in virtual environments: How do VR users process sensory stimuli? How do the mechanisms of central vision and peripheral vision affect our sense of presence in virtual space? Which of the stimuli mechanisms involved in neuronal processing should be taken into account—intensity, contrast, temporal frequency, or spatial frequency? Yet, designing reliable solutions that will stand the test of time requires more than ergonomic testing, though these tests too are necessary. Likewise, the issue of "presence" in virtual environments needs to be analysed, although this question on its own will not provide sufficient data for the development of effective new solutions. Our hope is that this book will incite neuroscientists to conduct in-depth methodical research on user response in virtual environments.

As to ensuring the appropriate development of VR applications, in terms of user-comfort and health, the behaviour of users immersed in virtual environments must be better understood. The evaluation of VR applications via user testing will therefore play a key role in guaranteeing captivating experiences that will be wholly endorsed by consumers. As previously noted, the field of virtual reality is still too young to have its own set of pre-existing norms, a lack of perspective which makes it difficult to imagine new forms of VR experiments. Even so, the user must not be considered as a mere "black box" with input and output sources. The aim of this book is to provide designers with new insights into the challenges of virtual reality.

Who knows, perhaps one day VR applications will depend on "virtuographers", just as 3D movies today rely on stereographers.

Foreseeable *culture changes* in video game and VR video design are already forcing designers to take a broader approach to questions of user immersion and interaction involving a growing number of interfacing variations, not all of which will become VE standards. What we do know is that the sacrosanct computer, with its joystick and console, is a thing of the past. A greater variety of interfacing technologies, both hardware and software, can be used to offer a wide range of sensorimotor and cognitive activities; Kinect, as mentioned earlier, is one such example of the revolution underway. But old habits die hard, and users are still reluctant to get out of their chairs and onto their feet—to adopt the erect posture of our ancestor, the Australopithecus, that is! Will the arrival of affordable VR headsets and the potential of virtual reality systems ultimately get players to stand up and challenge the classic video game paradigm? Of course, the use of VR headsets is in no way a prerequisite of video game development. It is, however, a visual interface offering the general public unprecedented opportunities for those recreational and creative activities.

Virtual reality is a science and technology that attempts to replicate the experience of the real world. So, future VR users will expect virtual world interaction to be modelled after interactions already familiar to them in the real world. It will be up to future VR professionals to provide the public with convincing alternative forms. Naturally, developers are free to use familiar real-world scenarios, but why stop there? Users could experience interactions beyond naturalistic virtual environments, which could involve a whole other set of non-realistic VBPs (Virtual Behavioural Primitives) developed for symbolic or imaginary worlds.

With the introduction of film, more than a century ago, the boundaries of media were redesigned. Yet this technical innovation hardly made a dent in the health of viewers—other than that alleged moment of panic when a life-size train seemed to be rushing straight at the audience! Next came 3D cinema making objects jump out of the screen – virtually – and causing a new and strange sense of visual trouble, an initial sensorimotor discrepancy. With today's VR applications viewers have become actual participants and, within the context of the "perception, decision, action" loop, they are being bombarded with disturbing sensorimotor discrepancies from all sides.

But what should we do about it, sit around and wait for VR headset designers and manufacturers to come up with all the answers? The fabrication of high-quality products suited to human capacities may be essential, but VR application developers and designers, whose work we have already examined in detail, are not the only masters of the game.

Indeed, virtual reality may be an "art form" with a number of key "ingredients", but the secret to successful VR applications is creativity. In other words, we must take a creative approach to interfaces and interfacing, to real-world and non-realistic VBPs, to activities that may or may not be true-to-life, to worlds that may be spatially, temporally and physically naturalistic, or that, then again, may not be...

14.2 PERSPECTIVES

Predicting the future is a tricky business, especially when it comes to innovative technologies and novel activities such as the ones being initiated by virtual reality. It is useful nonetheless to look back on certain predictions of old. For example, twenty years ago a number of journalists were very excited about the prospects of the emerging field: "Have you ever travelled in virtual reality? Those who have, have come back astounded: They have seen the Future!" Science et Vie, No. 990, June 1993[1]. It is important to distinguish between dream and reality without being overly influenced by science fiction, however. I have often been asked about the similarities between virtual reality and such famous science fiction movies as The Matrix, Minority Report or Avatar. It is the role of VR specialists to educate the public on what can – and cannot be – achieved with virtual reality.

It has already taken us some twenty years to master virtual reality systems that can be utilised for professional applications on a daily basis. Gradually, such applications are becoming more commonplace, although certain sectors have yet to take full advantage of the potential of VR. Yet the low-quality expensive headsets of yesterday have made way, at last, for today's more affordable versions, qualified by some as satisfactory. The question is whether it will take yet another two decades, or more, for quality VR headsets to finally enter the mainstream market.

While it is still relatively easy to predict the material evolution VR technologies, it is much harder to guess their future use, since the consumer is central to the question. How much longer will we have to wait—ten years, perhaps twenty or more—before VR experts and social science researchers will have acquired the experience and perspective needed to fully respond to such issues?

We must not let ourselves be influenced by the rapid evolution of a certain number of recent technologies, despite the enormous financial stakes. On the one hand, we have the early mobile phone, which rapidly evolved from an expensive unwieldy device with low autonomy to the current smartphone owned by one and all; on the other hand, there is the failure of inexpensive 3D television sets, despite certain predictions that 3D was the future of television programming! Yet this marginal success of 3D television usage was predicted by many in that it only responds to a secondary need of human vision, as previously mentioned. Though rendered with skill and creativity in 3D movies, the need to experience 3D film is tiny, compared to the pressing urge to communicate, on a daily basis, via smartphones. The same reasoning applies to the

---

[1]p118 – excerpt "La réalité virtuelle", Bernard Jolivalt, Que sais-je, 1995, ISBN 2-13-047290-7

foreseeable failure of Google Glass: users have no need for augmented reality on a daily basis.

Likewise, since the need for VR headsets is not as great as the need for smartphones, we are not expecting to see a run on the new products about to enter the market, knowing in addition that, unlike smartphones, these headsets can cause feelings of uneasiness or nausea. The subject of "virtual reality for the general public" has been widely covered by online media, with mixed reviews. Basically, it comes down to knowing whether the designers of future VR headset applications will be able to deliver immersion experiments in VEs with consistently positive user feedback. Entry level VR headset prices remain relatively high: over 700 hundred euros when you factor in recommended accessories. But prices may well drop, as they did for smartphones, if there is a massive demand for these devices in the future.

Fresh VR headset perspectives are offered by 360-degree video, beyond the ubiquitous family photo or video, though its utilisation is a good deal more restrictive. It remains to be seen whether such prospects will tempt the average consumer to acquire VR headsets, however. Yet coupled with consumer interest in innovative gaming experiences, there may be a strong enough demand to create a mass market, though one restricted to the financially affluent. Surveys have been conducted to determine potential demand for VR home applications. In one such survey, the Room Scale VR Survey[2], 69% of those who responded said they would be unwilling to move their computer to another part of the house to be able use an application requiring a VR headset; 42% reported they would be unwilling to upgrade their current computer to the more powerful hardware needed for VR applications. The future of VR headsets is more promising for devices designed for augmented reality (see Chapter 7), in view of the fact that AR headsets cause fewer unpleasant sensorimotor discrepancies, considered a major drawback by a significant portion of the population. However, these products too have yet to enter the market.

So VR headsets will continue being used for professional applications which will benefit from the latest technological developments, if no other visual interface adequately meets the sector's needs. The digital arts, too, will benefit from VR headset evolution, virtual environments allowing new forms of symbolic and imaginary works to be explored. But these uses will represent only a small part of the VR headset market.

The lion's share of the market will go to gaming, VR video and leisure activities, if the sensorimotor and cognitive capacities of individual players are convincingly accounted for.

---

[2] See https://www.surveymonkey.com/results/SM-YCHJCG32/

References

Airbus (2012) Available from: http://www.airbus.com/newsevents/news-events-single/detail/virtual-reality-supports-the-a350-xwbs-design-and-development/

Alfano, P.L. & Michel, G.F. (1990) Restricting the field of view: Perceptual and performance effects *Perceptual and Motor Skills*, 70, 35–45.

Allen, R.W., Rosenthal, T.J. & Cook, M.L. (2011) A short history of driving simulators. In: Fisher, D.L., Rizzo, M., Caird, J.K. & Lee, J.D. (eds.) *Handbook of Driving Simulation for Engineering, Medicine and Psychology*. Boca Raton, FL, CRC/Press Taylor and Francis. Chapter 2, pp. 2.1–2.16.

Angelaki, D.E. (2004) Eyes on target: What neurons must do for the vestibule-ocular reflex during linear motion? *Journal of Neurophysiology*, 92 (1), 20–35.

Arnaldi, B., Fuchs, P. & Tisseau J. (2003) *Chapitre 1 du volume 1 du traité de la réalité virtuelle*. Les Presses des Mines.

Barthou, A., Kemeny, A., Reymond, G., Merienne, F. & Berthoz, A. (2010) Driver trust and reliance on a navigation system: Effect of graphical display. In: Kemeny, A., Mérienne, F. & Espié, S. (eds.) *Trends in Driving Simulation Design and Experiments. Les Collections de l'INRETS*. pp. 199–208.

Batras, D., Guez, J., Jégo, J.-F. & Tramus, M.-H. (2016) *A Virtual Reality Agent-Based Platform for Improvisation Between Real and Virtual Actors Using Gestures*. Laval, France, ACM VRIC.

Berger, P. (2016) *DICCAN (Digital Creation Critical Analysis): The Dictionary of Digital Art*. Available from: http://diccan.com/ [Online, Accessed 12th September 2016].

Berthoz, A. (2000) *The Brain's Sense of Movement* Cambridge, MA, Harvard University Press. ISBN-10: 0674801091.

Berthoz, A. (2002) *The Brain's Sense of Movement*. Cambridge, MA, Harvard University Press.

Berthoz, A. (2008) *The Physiology and Phenomenology of Action*. Oxford, OUP Oxford. ISBN-10: 0199547882.

Bourriaud, N. (1998) *L'esthétique Relationnelle*. Les Presses du réel.

Bouvier, P. (2009) *La Présence En Réalité Virtuelle, Une Approche Centrée Utilisateur*. Thesis. University Paris-Est, dir. Gilles Bertrand.

Bradshaw, M.F., Parton, A.D. & Glennerster, A. (2000) The task-dependent use of binocular disparity and motion parallax information *Vision Research*, 40, 3725–3734.

Bret, M., Tramus, M.-H. & Berthoz, A. (2005) Interacting with an intelligent dancing figure: Artistic experiments at the crossroad between art and cognitive science. *Leonardo for the Art Sciences and Technology*, 38, 46–53.

Burns, P.C. & Saluäär, D. (1999) Intersections between driving in reality and virtual reality. In: *Proceedings of the Driving Simulation Conference 1999*. pp. 155–164.

Coates, N., Ehrette, M., Blackham, G., Heidet, A. & Kemeny, A. (2002) Head-mounted display in driving simulation applications in CARDS. In: *Proceedings of the Driving Simulation Conference 2002, Paris, France*. pp. 33–43.

Coello, Y. & Grealy, M.A. (1997) Effect of size and frame of visual field on the accuracy of an aiming movement. *Perception*, 26, 287–300.

Coleridge, S.T. (1772–1934) *Biographie Literaria*. Available from: http://www.gutenberg.org/files/6081/6081-h/6081-h.htm#link2H_4_0002 [e-Book 2004: Accessed 6th February 2015].

Colombet, F., Kemeny, A. & George, P. (2016) Motion sickness comparison between a CAVE environment and a HMD. In: *Proceedings of the Driving Simulation Conference Europe 2016 VR, Paris, France*. pp. 201–208.

Couchot, E. (1998) *La Technologie Dans L'art: De La Photographie à La Réalité Virtuelle*. Nîmes, Jacqueline Chambon.

Cruz-Neira, C., Sandin, D.J., DeFanti, T.A., Kenyon, R.V. & Hart, J.C. (1992) The CAVE: Audio visual experience automatic virtual environment. *Communications of the ACM*, 35 (6), 64–72.

Cruz-Neira, C., Sandin, D.J. & DeFanti, T.A. (1993) Surround-screen projection-based virtual reality: The design and implementation of the CAVE. In: *SIGGRAPH'93: Proceedings of the 20th Annual Conference on Computer Graphics and Interactive Techniques*. pp. 135–142.

Dagdelen, M., Reymond, G. & Kemeny, A. (2002) Analysis of the visual compensation in the Renault driving Simulator. In: *Proceedings of the Driving Simulation Conference, Paris, France, 2002*. pp. 109–119.

DeFanti, T.A., Acevedo, D., Ainsworth, R.A., *et al.* (2010) The future of the CAVE. *Central European Journal of Engineering*, 1, 16–37.

Draper, M.H., Viire, E.S., Furness, T.A. & Gawron, V.J. (2001) Effects of image scale and system time delay on simulator sickness with head-coupled virtual environments. *Human Factors*, 43 (1), 129–146.

Drosdol, J. & Panik, F. (1985) *The Daimler-Benz Driving Simulator, A Tool for Vehicle Development*. SAE paper, No. 850334.

Evangelakos, D. & Mara, M. (2016) Extended TimeWarp latency compensation for virtual reality. In: *Proceedings of the 20th ACM SIGGRAPH Symposium on Interactive 3D Graphics and Games (I3D '16)*. New York, NY, ACM. pp. 193–194.

Fang, Z., Reymond, G. & Kemeny, A. (2011) Performance identification and compensation of simulator motion cueing delays. *Journal of Computing and Information Science in Engineering, Special Issue in Driving Simulation*, 11 (4), 1003-1–1003-4.

Filliard, N., Reymond, G., Kemeny, A. & Berthoz, A. (2012) Motion parallax in immersive cylindrical display systems In: *Proceedings of SPIE*, Vol. 8289, 24–25 January 2012.

Flanagan, M.B., May, J.G. & Tobie, T.G. (2005) Sex differences in tolerance to visually induced motion sickness. *Aviation Space and Environmental Medicine*, 76 (7), 642–646.

Foreman, N., Sandamas, G. & Newson, D. (August 2004) Distance underestimation in virtual space is sensitive to gender but not activity-passivity or mode of interaction. *Cyberpsychology & Behavior*, 7 (4), 451–457.

Fuchs, P. (1996) *Les interfaces de la réalité virtuelle*. Les Presses des Mines. ISBN: 2-9509954-0-3.

Fuchs, P., Ernadotte, D., Maman, D., Laurgeau, C. & Bordas, J. (1995) Téléprésence virtuelle stéréoscopique. In: *Interface des Mondes réels et virtuels*. France, Montpellier. pp. 77–91.

Fuchs, P., Guitton, P. & Moreau, G. (2011) *Virtual Reality: Concepts and Technologies*. CRC Press. ISBN 9780415684712.

Fuchs, P., Nashashibi, F. & Lourdeaux, D. (1999) A theoretical approach of the design and evaluation of a virtual reality device. In: *Virtual Reality and Prototyping'99*. Laval, France. pp. 11–20.

Fuchs, P., Moreau, G., Berthoz, A. & Vercher, J.-L. (2006) *Le traité de la réalité virtuelle volume 1 – L'Homme et l'environnement virtuel*. Presses de l'École des Mines.

George, B., Kemeny, A., Colombet, F., Merienne, F., Chardonnet, J. & Thouvenin, I. (2014) Evaluation of smartphone-based interaction techniques in a CAVE in the context of immersive digital project review. In: *Proceedings of Conference: IS&T/SPIE Electronic Imaging, The Engineering Reality of Virtual Reality, Vol. 9012, 02 February, 2014*.

Gibson, J. (1966) *The Senses Considered as a Perceptual System*. Boston, MA, Houghton Mifflin.

Gibson, J.J. (1979) *The Ecological Approach to Visual Perception*. Hillsdale, MI, Lawrence Erlbaum Associates.

Glennerster, A., Tcheang, L., Gilson, S.J., Fitzgibbon, A.W. & Parker, A.J. (2006) Humans ignore motion and stereo cues in favor of a fictional stable world. *Current Biology*, 16 (4), 428–432.

Gogel, W.C. & Tietz, J.D. (1973) Absolute motion parallax and the specific distance tendency. *Perception & Psychophysics*, 13 (2), 284–292.

Guez, J. (2013) *De L'interaction à La Présence – Un Art Qui Se Vit*. Proteus, Le spectateur face à l'art interactif 6.

Guez, J. (2015) *Illusions Entre Le Réel et Le Virtuel (IRV) Comme Nouvelles Formes Artistiques: Présence et émerveillement*. Thesis, Esthétique, Science et Technologie des Arts, University Paris 8, dir. Marie-Hélène Tramus.

Hagen, M.A., Jones, R.K. & Reed, E.S. (1978) On a neglected variable in theories of pictorial perception: Truncation of the visual field. *Perception & Phychopsychics*, 23, 326–330.

Hale, K.S. & Stanney, K.M. (2006) Effects of low stereo acuity on performance, presence and sickness within a virtual environment. *Applied Ergonomics*, 37, 329–339.

Harm, D.L. (2002) Motion sickness neurophysiology, physiological correlates, and treatment. In: Stanney, K.M. (ed.) *Handbook of Virtual Environments: Design, Implementation, and Applications*. Mahwah, IEA. pp. 637–661.

Harris, L.R., Carnevale, M.J., D'Amour, S., Fraser, L.E., Harrar, V., Hoover, A.E.N., *et al.* (2015) How our body influences our perception of the world. *Frontiers in Psychology*, 6, 819.

Hoffman, D.M., Girschick, A.R., Akeley, K. & Banks, M.S. (2008) Vergence accommodation conflicts hinder visual performance and cause visual fatigue. *Journal of Vision*, 8 (3), 1–30.

Howard, I.P. & Rogers, B.J. (1995) *Binocular Vision and Stereopsis*. New York, NY, Oxford University Press.

Huon, A. (2015) *Interview sur la réalité virtuelle avec Jenova Chen, créateur de Journey*. Available from: http://www.etr.fr/interview/2310-interview-sur-la-realite-virtuelle-avec-jenova-chen-createur-de-journey.html [Online: Accessed 12th September 2016].

Jamson, A.H. (2000) Driving simulation validity. In: *Proceedings of the Driving Simulation Conference, Paris, France, 2000*. pp. 57–64.

Julesz, B. (1971) *Foundation of Cyclopean Perception*. Chicago, IL, University of Chicago Press.

Kemeny, A. (1987) Synthèse d'images fixes et animées. In: *Techniques de l'ingénieur*, E 5 530, 1–21.

Kemeny, A. (2000) Simulation and perception of movement. In: *Proceedings of the Driving Simulation Conference, Paris, 2000*. pp. 13–22.

Kemeny, A. (2001) Recent developments in visuo-vestibular restitution of self-motion in driving simulation. In: *Proceedings of the Driving Simulation Conference, Sophia Antipolis, France, 2001.* pp. 15–18.

Kemeny, A. (2009) Driving simulation for virtual testing and perception studies. In: *Proceedings of the Driving Simulation Conference Europe 2009, Monte Carlo.* pp. 15–23.

Kemeny, A. (2014) From driving simulation to virtual reality. In: *Proceedings of the VRIC'14 Virtual Reality International Conference, Laval, France, 2014, Art. No. 32.*

Kemeny, A. & Panerai, F. (2003) Evaluating perception in driving simulation experiments. *Trends in Cognitive Sciences,* 7, 31–37.

Kemeny, A., Combe, E. & Posselt, J. (2008) Perception of size in vehicle architecture studies. In: *Proceedings of the 5th Intuition International Conference, Torino, Italy*

Kemeny, A., Aykent, B., Yang, Z. & Merienne, F. (2014) Simulation sickness comparison between a limited field of view virtual reality head mounted display and a medium range field of view static ecological driving simulator (ECO2). In: *Driving Simulation Conference Europe 2014 Proceedings, Sep 2014, Paris, France.* Society for Modeling & Simulation International. pp. 65–71.

Kemeny, A. Colombet, F. & Denoual, T. (March 2015) How to avoid simulation sickness in virtual environments during user displacement. In: *Proceedings of Conference: IS&T/SPIE Electronic Imaging, The Engineering Reality of Virtual Reality,* Vol. 9392. pp. 939206.1–939206.9.

Kennedy, R.S., Lane, N.E., Berbaum, K.S. & Lilienthal, M.G. (1993) Simulator sickness questionnaire: An enhanced method for quantifying simulator sickness. *The International Journal of Aviation Psychology,* 3 (3), 203–220.

Kennedy, R.S., Lane, N.E., Grizzard, M.C., Stanney, K.M., Kingdon, K. & Lanham, S. (2001) Use of a motion history questionnaire to predict simulator sickness. In: *Proceedings of the Driving Simulation Conference, Sophia Antipolis, France, 2001.* pp. 79–89.

Knapp, J.M. & Loomis, J.M. (2004) Limited field of view of head mounted displays is not the cause of distance underestimation in virtual environments. *Presence: Teleoperators & Virtual Environments,* 13, 572–577.

Konstantina, K., Normand, J.M., Sanchez-Vives, M.V. & Slater, M. (2012) Extending body space in immersive virtual reality: A very long arm illusion. *PLoS One,* 7 (7), e40867.

Lanier, J. (1990) Life in the data-cloud. *Mondo 2000,* 2, 4–51

Lécuyer, A., Coquillart, S., Kheddar, A., Richard, P. & Coiffet, P. (2000) Pseudo-haptic feedback: Can isometric input devices simulate force feedback? In: *Proceedings of the IEEE International Conference on Virtual Reality.*

Lécuyer, A., Burkhardt, J.M. & Etienne, L. (2004) Feeling bumps and holes without a haptic interface: The perception of pseudo-haptic textures. In: *ACM Conference in Human Factors in Computing Systems (ACM SIGCHI'04), April 24–29, Vienna, Austria.*

Lepecq, J., Bringoux, L., Pergandi, J., Coyle, T. & Mestre, D. (2009) Afforded actions as a behavioral assessment of physical presence in virtual environments. *Virtual Reality,* 13, 141–151.

Leroy, L., Fuchs, P., Paljic, A. & Moreau, G. (2009) Some experiments about shape perception in stereoscopic displays. In: *IS&T/SPIE Symposium on Electronic Imaging, Number 7237-45, 18-22 January 2009, San Jose, California USA.*

Loomis, J.M. & Knapp, J.M. (2003) Visual perception of egocentric distance in real and virtual environments. In: Hettinger, L.J. & Haas, M.W. (eds.) *Virtual and Adaptive Environments.* Mahwah, NJ, Erlbaum. pp. 21–46.

Loomis, J.M., Blascovich, J.J. & Beall, A.C. (1999) Immersive virtual environment technology as a basic research tool in psychology. *Behavior Research Methods, Instruments and Computers,* 31, 557–564.

Lorissin, J. (2010) Réalité virtuelle dans l'industrie – Dévelopement des produits et des processus *Tecniques de l'ingénieur*, 5, 965.

Mestre, D.R. (2016) Perceptual calibration in virtual reality applications. *The Engineering Reality of Virtual Reality 2016*, 6, 1–6.

Mestre, D.R. & Fuchs, P. (2006) Immersion et Présence. In: Fuchs, P., Moreau, G., Berthoz, A. & Vercher, J.L. (eds.) *Traité de la Réalité Virtuelle, Volume 1 L'homme et l'environnement virtuel*. Paris, Ecole des Mines de Paris. pp. 309–338.

Mestre, D.R. & Vercher, J.L. (2011a) Chapter 5, Immersion and presence. In: Fuchs, P., Moreau, G. & Guitton, P. (eds.) *Virtual Reality: Concepts and Technologies*. Boca Raton, FL, CRC Press. 432 pp. ISBN: 9780415684712.

Mestre, D.R. & Vercher, J.L. (2011b) Interaction between virtual reality and behavioral sciences. In: Fuchs, P., Moreau, G. & Guitton, P. (eds.) *Virtual Reality: Concepts and Technologies*. New York, NY, CRC Press. pp. 81–91.

Mestre, D.R., Louison, C. & Ferlay, F. (2016) The contribution of a virtual self and vibrotactile feedback to walking through virtual apertures. *Human-Computer Interaction. Interaction Platforms and Techniques*, Vol. 9732 of the series Lecture Notes in Computer Science. pp. 222–232.

Milk, C. (2015) How virtual reality can create the ultimate empathy machine. In: *Conference TED2015*. Available from: https://www.ted.com/talks/chris_milk_how_virtual_reality_can_create_the_ultimate_empathy_machine?language=en [Online video: Accessed 12th September 2016].

Mohellebi, H., Espié, S., Arioui, H., Amouri, A. & Kheddar, A. (2004) Low cost motion platform for driving simulator. In: *ICMA: International Conference on Machine Automation, Japan. 5th*. pp. 271–277.

Nannipieri, O. & Fuchs, P. (2009) Pour en finir avec la réalité: une approche socio-constructiviste de la réalité virtuelle. *Revue des Interactions Humaines Médiatisées*, 10 (1), 83–100.

Nannipieri, O., Muratore, I., Mestre, D. & Lepecq, J.-C. (2014) *La Présence Dans La Réalité Virtuelle: Quand La Frontière Se Fait Passage*. Frontières Numériques.

Nelson, W.T., Roe, M.M., Bolia, R.S. & Morley, R.M. (2000) Assessing simulator sickness in a see-through HMD: Effects of time delay, time on task, and task complexity. In: *Proc. IMAGE Conference ASC-00-1047*.

Nordmark, S. (1994) Driving simulators, trends and experiences. In: *Proceedings of the Driving Simulation Conference, Real Time Systems, Paris*. pp. 5–13.

Paillé, D., Kemeny, A. & Berthoz, A. (2005) Stereoscopic stimuli are not used in absolute distance evaluation to proximal objects in multi-cue virtual environment. In: *Proceedings of SPIE*, Vol. 5664. pp. 596–605.

Panerai, F., Droulez, J., Kelada, J.-M., Kemeny, A., Balligand, E. & Favre, B. (2001) Speed and safety distance control in truck driving: Comparison of simulation and real-world environment. In: *Proceedings of the Driving Simulation Conference, Sophia Antipolis, France, 2001*. pp. 21–32.

Panerai, F., Cornilleau-Peres, V. & Droulez, J. (2002) Contribution of extraretinal signals to the scaling of object distance during self-motion. *Perception & Psychophysics*, 64, 717–731.

Perrin, J. (1998) *Profondeur et binocularité: algorithmie, étude psychophysique et intérêt pour l'ergonomie des interfaces stéréoscopiques*. PhD Thesis. Paris, Ecole des Mines ParisTech.

Perrin, J., Fuchs, P., Roumes, C. & Perret, F. (1998) Improvement of stereoscopic comfort through control of disparity and spatial frequency content. In: *Visual Information Processing VII, Volume 3387 of Proceedings of SPIE*.

Piaget, J. & Chomsky, N. (1979) *Théories du langage, théories de l'apprentissage*. Seuil.

Popper, F. (1985) *Art, Action et Participation: L'artiste et La Créativité Aujourd'hui.* Klincksieck.

Rabardel, P. (1995) *Les hommes et les technologies, approche cognitive des instruments contemporains.* Armand Colin. ISBN: 2-200-21569-X.

Reason, J.T. & Brand, J. (1975) *Motion Sickness.* London, Academic Press.

Riccio, G.E. & Stoffregen, T.A. (1991) An ecological theory of motion sickness and postural instability. *Ecological Psychology*, 3, 195–240.

Rock, I. & Harris, C.H. (1967) Vision and touch. *Scientific American*, 2, 8.

Rogers, B. & Graham, M. (1979) Motion parallax as an independent cue for depth perception. *Perception*, 8, 125–134.

Schiller, V., *et al.* (1997) Car research using virtual reality at Daimler-Benz. In: *Proceedings of the Driving Simulation Conference 1997, Lyon, France.* pp. 35–44.

Shermann, W.R. & Craig, A.B. (2003) *Understanding Virtual Reality.* San Francisco, CA, Morgan Kaufmann.

Sherrington, C.S. (1906) *The Integrative Action of the Nervous System.* New Haven, CT, Yale University Press (1947 ed.).

Shibata, T., Kim, J., Hoffman, D.M. & Banks, M.S. (2011) The zone of comfort: Predicting visual discomfort with stereo displays. *Journal of Vision*, 11 (8), 1–29.

Siegler, I., Reymond, G., Kemeny, A. & Berthoz, A. (2001) Sensorimotor integration in a driving simulator: Contributions of motion cueing in elementary driving tasks. In: *Proceedings of the Driving Simulation Conference, Sophia Antipolis, France, 2001.* pp. 21–32.

Slater, M. (2003) A note on presence terminology. *Emotion*, 3, 1–5.

Slater, M. (2009) Place illusion and plausibility can lead to realistic behaviour in immersive virtual environments. *Philosophical Transactions of the Royal Society of London. Series B, Biological Sciences*, 364 (1535), 3549–3557.

Sohier, R. (2016) *L'expérience émersive du jeu video.* Implications philosophiques. Available from: http://www.implications-philosophiques.org/actualite/une/lexperience- emersive-du-jeu-video/ [Online: Accessed 12th September 2016].

Stanney, K.M. & Hash, P. (1998) Locus of user-initiated control in virtual environments: Influences on cybersickness. *Presence*, 7 (5), 447–459.

Stanney, K.M., Kennedy, R.S. & Kingdon, K. (2002a) Virtual environment usage protocols. In: Stanney, K.M. (ed.) *Handbook of Virtual Environments: Design, Implementation, and Applications.* Mahwah, IEA. pp. 721–730.

Stanney, K.M., Kingdon, K.S., Graeber, D. & Kennedy, R.S. (2002b) Human performance in immersive virtual environments: Effects of exposure duration, user control, and scene complexity. *Human performance*, 15, 339–366.

Stefanucci, J.K., Sarah, H., Creem-Regehr, S.H. Thompson W.B. & Mohler, B.J. (November 2015) Effect of display technology on perceived scale of space. *Human Factors*, 57 (7), 1235–1247.

Sutherland, I.E. (1965) The ultimate display. In: *Proceedings of the Information Processing Techniques Congress.* pp. 506–508.

Sutherland, I.E. (1968) *A Head-Mounted Three Dimensional Display.* Washington, DC, FJCC Thompson Books. pp. 757–764.

Szymanski, M. (2002) *An Introduction to Curved Screen Displays.* White paper. vRco, Available from: www.vrco.com

Tarr, M.J. & Warren, W.H. (2002) Virtual reality in behavioral neuroscience and beyond. *Nature Neuroscience*, 5, 1089–1092.

Tramus, M.-H. (2001) *Recherches, Expérimentations et Créations Dans Les Arts Numériques: Interactivité, Acteurs Virtuels.* ASR. University Paris 8.

Treisman, M. (1977) Motion sickness: An evolutionary hypothesis. *Science*, 197, 493–495.

Trotter, Y. (1995) Bases neuronales de la perception visuelle chez le primate. *Journal français d'orthoptique*, 27, 9–20.

Valyus, N. (1962) *Stereoscopy*. London, Focal Press.

Vermersch, P. (2012) *Explicitation et Phénoménologie: Vers Une Psychophénoménologie*. PUF.

Voillequin, T. (2006) First steps of Haptics at PSA Peugeot Citroën. Special session of industrial applications of force feed-back, VRIC'06, LAVAL.

Warren, W. & Whang, S. (1987) Visual guidance of walking through apertures: Body-scaled information for affordances. *Journal of Experimental Psychology: Human Perception and Performance*, 13, 371–383

Watt, S.J., Bradshaw, M.F. & Rushton, S.K. (2000) Field of view affects reaching, not grasping. *Experimental Brain Research*, 135, 411–416.

Wiederhold, B. & Bouchard, S. (2014) *Advances in Virtual Reality and Anxiety Disorders*. Boston, MA, Springer US.

Yeh, Y. & Silverstein, L. (1990) Limits of fusion and depth judgment in stereoscopic color displays. *The Human Factors Society*, 32, 45–60.

ARTWORKS

Auxietre, B. (2015a) *Playhead*. Music rhythm Virtual Reality experience, HMD, Innerspace.

Auxietre, B. (2015b) *The 5th Sleep*. Virtual Reality installation, HMD, Innerspace.

Batras, D., Jégo, J.-F., Guez, J. & Tramus, M.-H. (2015) *InterACTE*. Virtual reality installation, HMD, Ars Electronica. Available from: http://cigale.labex-arts-h2h.fr [Online: Accessed 19th August 2015].

BeAnotherLab Collective (2015) *The Machine to Be Another*. Live video Installation, HMD.

Benayoun, M. (1998) *World Skin*. Virtual Reality installation, CAVE, Ars Electronica Center.

Boissier, J.-L. (1986) *Pékin pour mémoire*. Interactive installation. Video projection and table with buttons.

Bret, M. & Tramus, M.-H. (2000–2006) *la Funambule virtuelle*. Second interactivity installation, Virtual Reality installation.

Chabalier, C., Stamatiadi, V., Turlet, J., Sivinia, A., Morisset, T., Valette, E., Bisbicki, V., Jego, J.-F., Guez, J., Batras, D. & Danet, C. (2016) *Cassandre-Matériaux*. Digital theater play, Video projection and HMD, Theater: la Commune, Aubervilliers, France.

Davies, C. (1995) *Osmose*. Virtual Reality installation, HMD.

Duchamp, M. (1920) *Rotary Glass Plates*. Installation.

Fleuryfontaine (2014) *I Need a Haircut*. Virtual Reality installation, Galerie YGREC.

Guez, J., Bertinet, G., Wagrez, K. & Costes, F. (2015a) *Lab'Surd: le LABoratoire de la SURvirtualité*. Virtual Reality installation, Ars Electronica.

Guez, J., Jégo, J.-F., Wagrez, K., Gilly-Poitou, G., Loosveld, J., Moussier, C. & Rivoire, D. (2015b) *Liber*. Virtual Reality artistic game, Bibliothèque Publique d'Information (BPI), Centre Pompidou, France.

Kruger, M. (1970) *Videoplace*. Artificial Reality installation.

Lin, J.-Y. (2016) *Sensation World*. Interactive installation with live video and 3D printed objects.

Spiers, J., Eveillard, L. & Dervieux, F. (2015) Panopticon. In: *VRJam 2015*.

Tinguely, J. & de Saint Phalle, N. (1969–1994) *Le Cyclop*. Immersive sculpture, Milly-la-Forêt France.

Subject index

Milton Keynes UK
Ingram Content Group UK Ltd.
UKHW051854071024
449327UK00025B/1949